# EMC of Analog Integrated Circuits

**ANALOG CIRCUITS AND SIGNAL PROCESSING SERIES**

*Consulting Editor*: **Mohammed Ismail.** ***Ohio State University***

For other titles published in this series, go to
www.springer.com/series/7381

Jean-Michel Redouté • Michiel Steyaert

# EMC of Analog Integrated Circuits

Springer

Jean-Michel Redouté
Dept. Electrical Engineering (ESAT)
Katholieke Universiteit Leuven
Kasteelpark Arenberg 10
3001 Leuven, Belgium
Jean-Michel.Redoute@esat.kuleuven.be

Michiel Steyaert
Dept. Electrical Engineering (ESAT)
Katholieke Universiteit Leuven
Kasteelpark Arenberg 10
3001 Leuven, Belgium
michiel.steyaert@esat.kuleuven.be

*Series Editor*:
Mohammed Ismail
205 Dreese Laboratory
Department of Electrical Engineering
The Ohio State University
2015 Neil Avenue
Columbus, OH 43210, USA

ISBN 978-94-007-3088-5 e-ISBN 978-90-481-3230-0
DOI 10.1007/978-90-481-3230-0
Springer Dordrecht Heidelberg London New York


Softcover reprint of the hardcover 1st edition 2010


*Cover design*: eStudio Calamar S.L.

Printed on acid-free paper

Springer is part of Springer Science+Business Media (www.springer.com)

# Preface

Environmental electromagnetic pollution has drastically increased over the last decades. The omnipresence of (wireless) communication systems, new and various electronic appliances and the use of ever increasing frequencies, all contribute to a noisy electromagnetic environment which acts detrimentally on sensitive electronic equipment. Integrated circuits constitute the beating heart of almost any given electronic system nowadays: luckily, owing to their small sizes, they are not easily disturbed by radiated disturbances, because their tiny on-chip interconnections are much too small to function as effective antennas. However, the ultimate contribution comes from the conducted interferences which are present on the noisy and relatively long printed circuit board tracks, used to connect and interconnect the integrated circuits in question.

Aside from a polluted electromagnetic spectrum, integrated circuits must be able to operate satisfactorily while cohabiting harmoniously in the same appliance, and not generate intolerable levels of electromagnetic emission, while maintaining a sound immunity to potential electromagnetic disturbances. As different electronic systems are compactly integrated in the same apparatus, the parasitic electromagnetic coupling between these circuits sharing the same signal, power and ground lines, is a critical design parameter that can no longer be safely excluded from a product design flow. This dense integration level links the electromagnetic compatibility (EMC) issue of integrated circuits to the graceful coexistence between systems: as an example, Bluetooth, GSM and WiFi services have to coexist and operate in harmony within the crammed confinement of a modern mobile phone.

Distinct frequency allocations provide a shield against electromagnetic interferences by separating the signal spectra of different systems from each other: nevertheless, the intrinsic nonlinearity of active devices may cause the demodulation of interfering out of band signals, whereby spurious signals tend to appear in the frequency band of the exposed circuit. Furthermore, these out of band disturbances may induce a severe distortion of the wanted signals (which is certainly not recommended), and DC shift errors on sensitive nodes in the respective circuit, hereby possibly driving the latter out of its correct operation mode (which is even less acceptable). Analog circuits are in practice more easily disturbed than their digital counterparts, since they don't have the benefit of dealing with predefined levels which ensure an innate immunity to disturbances.

The objective of the research domain presented in this book is to improve the electromagnetic immunity of considered analog integrated circuits, so that they start to fail at relevantly higher conduction levels than before.

J.-M. Redouté and M. Steyaert

# Contents

# Chapter 1

# Introduction

*Medical technicians taking a heart-attack victim to the hospital in 1992 attached her to a monitor/defibrillator. Unfortunately, the heart machine shut down every time the technicians turned on their radio transmitter to ask for advice, and as a result the woman died. Analysis showed that the monitor unit had been exposed to exceptionally high fields because the ambulance roof had been changed from metal to fibreglass and fitted with a long-range radio antenna. The reduced shielding from the vehicle combined with the strong radiated signal proved to be too much for the equipment.*

—Quoted from [Arm07]

## 1 The pioneers of wireless communication

Since ancient times, people have been aware of the magnetic properties of the lodestone. According to Aristotle, Thales of Miletus attributed the attraction of the lodestone on iron to the fact that "it has a soul" [Ida04]. The earliest mention of the magnetic attraction of a needle appears in a Chinese work composed by Louen-heng between 20 and 100 AD, stating that "a magnetic stone attracts a needle" [Shu54]. In 1600, William Gilbert (who first used the adjective *electric* after the Greek word for amber, *electron*) published his work on magnetism and electricity ("De Magnete, Magneticisque Corporibus, et de Magno Magnete Tellure"), with which the modern history of both subjects begins [Whi60]. However, both electricity as well as magnetism were still considered to be separate entities. In 1820, the Danish scientist Hans Christian Ørsted, Professor of Natural Philosophy in Copenhagen, observed during a course of lectures on "Electricity, Galvanism and Magnetism" that a compass needle deflects from the magnetic north when the current in a nearby electric wire is switched on [Whi60]. After intensive research and diverse experiments, Ørsted, a strong believer in the unity of nature's forces, came to the conclusion that an electric current flowing through a conductor generates a magnetic

J.-M. Redouté, M. Steyaert, *EMC of Analog Integrated Circuits*,
Analog Circuits and Signal Processing,
DOI 10.1007/978-90-481-3230-0_1, © Springer Science+Business Media B.V. 2010

field which is circular around the conductor and whose intensity is directly proportional to the current itself. As a consequence, he established the unique relationship existing between magnetism and electricity, and the scientific discipline of electromagnetism was born out of his findings. The four Maxwell's equations, which were presented by their author in 1864, combine the relationships between electricity and magnetism in a very concise way: these equations combine the works of Faraday, Ampère, Gauss, Thomson and others, and elegantly merge the properties of electricity to those of magnetism in a united mathematical framework [Whi60]. From then on, it was just a matter of time before electromagnetism led to wireless and wireline communication.

Radio is considered as one of the most fabulous and wonderful inventions fathered at the end of the 19th century, leaning on the magic side since the scientists, engineers, inventors and pioneers having developed it could actually not fathom exactly how it worked at that time. An interesting and very entertaining account of the history of radio communication is presented eloquently in [Wei03]. Owing to the multitude of people who contributed to a larger or lesser extent to the invention of radio wave communication, it would be very unjust to credit just one as the real inventor of the radio, although some of them (like Lee the Forest) did not hesitate to monopolize this prestigious title for themselves. The numerous experiments, developed equipment and test setups all testify to the fact that wireless communication through radio waves is a shared invention: this fact can amongst others be appreciated in [Phi80], in which an interesting summary of early radio detectors is presented. However, some important milestones in the tumultuous history of radio can objectively be distinguished.

In 1888, Heinrich Rudolf Hertz developed the first wireless receiver and transmitter in his laboratory and measured the generated electric field strength as well as its polarity: he hereby demonstrated through experimentation that electromagnetic waves exist, and that they travel a certain distance through air [Whi60]. In 1893, after successfully winning the "war of currents" by establishing the indisputable superiority of the alternating current power system over the anterior DC power system supported by Thomas Edison [1], brilliant Nikola Tesla publicly demonstrated the principles of radio waves and filed his radio patent (US645576) in 1900. One year later, in 1901, young Guglielmo Marconi combined previous knowledge and basically without inventing anything new, successfully set up the first transatlantic radio communication between Saint John's (Newfoundland) and Poldhu (Great-Britain) by transmitting the Morse code of the letter 'S' (dot-dot-dot), hereby spanning a distance of approximately 3500 km. When presented with this fact, Nikola Tesla reacted dryly

[1] This fact accounts for the prevalence of the AC power system nowadays.

that Marconi was actually using seventeen of his patents. Suddenly, and quite against all odds, Marconi obtained the famous "four sevens" (British patent number 7777) patent on radio in 1904, after the US patent Office mysteriously reversed its decision of granting Tesla the patent of the radio which had been filed a few years earlier. Five years later, Marconi was even awarded the Nobel prize for physics (co-shared with Karl Ferdinand Braun) in 1909. This famous British "four-sevens" patent and its equivalents in other countries, as well as the acclaimed Nobel prize, triggered a set of unrelenting and very brutal patent legal disputes between Marconi and Tesla, often leading to arbitrary rulings which varied between the full nullification to the full acceptance of Marconi's radio patent. In 1943, a few months after Tesla's death, this legal onslaught was resolved in the United States when the US Supreme Court upheld Tesla's original radio patent (US645576), confirming the importance of prior research which had been conducted by Nikola Tesla, Oliver Lodge, John Stone Stone and others. However, this decision which finally recognized and credited the genius of Nikola Tesla was not fully guided by pure altruism: since the Marconi Company was suing the United States Government for the use of its patents during the first World War, the Court simply avoided this action by restoring the priority of Tesla's patent over Marconi.

Since these early radio days, the quantity of wireless communication devices has been continuingly increasing, and the amount of electromagnetic compatibility (EMC) problems between communication appliances coexisting in the same environment have been increasing simultaneously. It would be incorrect, however, to attribute the origin of electromagnetic compatibility issues solely to the presence of wireless communication systems.

## 2 Evolution of awareness of electromagnetic compatibility

In 1892, an edict issued by the German parliament and signed by Wilhelm II, Kaiser of Germany and King of Prussia [2], indicated that in the event electromagnetic disturbances perturb the correct operation of telegraph cables or telegraph equipment, the owner(s) of the appliance(s) that is (are) responsible of causing the disturbances should solve the problem and indemnify the owner(s) of the telegraph cables and telegraph equipment whom he has disturbed (Fig. 1.1). This edict is an important milestone in the history of EMC in the sense that it was the first known regulation to be adopted which attempted to make electrical appliances compatible with each other.

The first EMC problem between two different appliances that has been the subject of intensive studies and scientific investigations, is the radio receiver which is fitted in a car. The primary cause for this generated interference is the

[2] *Gesetz über das Telegraphenwesen des Deutschen Reichs, Reichsgesetzblatt S. 467.*

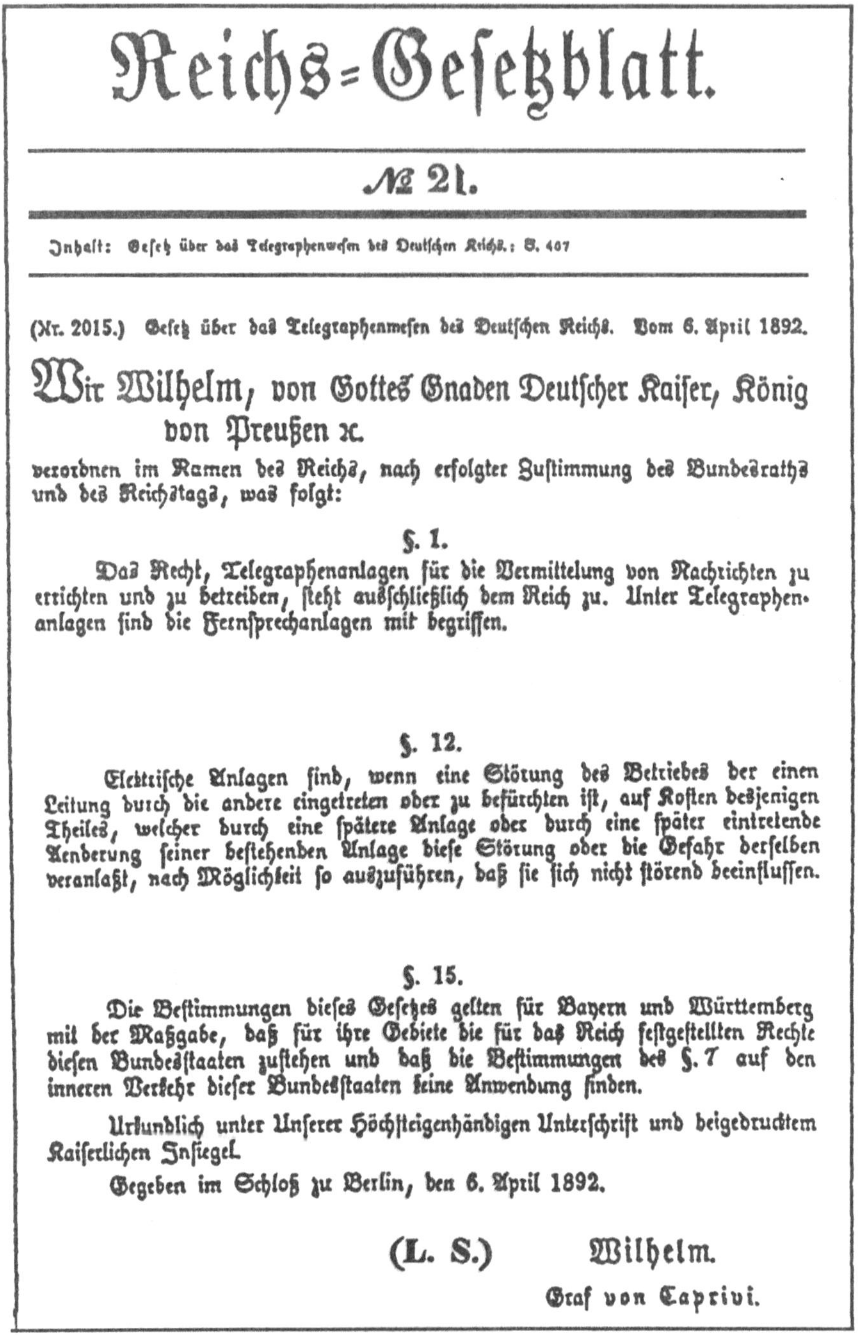

# Reichs-Gesetzblatt.

№ 21.

Inhalt: Gesetz über das Telegraphenwesen des Deutschen Reichs. S. 407

(Nr. 2015.) Gesetz über das Telegraphenwesen des Deutschen Reichs. Vom 6. April 1892.

Wir Wilhelm, von Gottes Gnaden Deutscher Kaiser, König von Preußen rc.

verordnen im Namen des Reichs, nach erfolgter Zustimmung des Bundesraths und des Reichstags, was folgt:

§. 1.

Das Recht, Telegraphenanlagen für die Vermittelung von Nachrichten zu errichten und zu betreiben, steht ausschließlich dem Reich zu. Unter Telegraphenanlagen sind die Fernsprechanlagen mit begriffen.

§. 12.

Elektrische Anlagen sind, wenn eine Störung des Betriebes der einen Leitung durch die andere eingetreten oder zu befürchten ist, auf Kosten desjenigen Theiles, welcher durch eine spätere Anlage oder durch eine später eintretende Aenderung seiner bestehenden Anlage diese Störung oder die Gefahr derselben veranlaßt, nach Möglichkeit so auszuführen, daß sie sich nicht störend beeinflussen.

§. 15.

Die Bestimmungen dieses Gesetzes gelten für Bayern und Württemberg mit der Maßgabe, daß für ihre Gebiete die für das Reich festgestellten Rechte diesen Bundesstaaten zustehen und daß die Bestimmungen des §. 7 auf den inneren Verkehr dieser Bundesstaaten keine Anwendung finden.

Urkundlich unter Unserer Höchsteigenhändigen Unterschrift und beigedrucktem Kaiserlichen Insiegel.

Gegeben im Schloß zu Berlin, den 6. April 1892.

(L. S.) Wilhelm.

Graf von Caprivi.

*Figure 1.1.* Excerpt of the edict issued by the German parliament, regulating electromagnetic disturbances in the telegraph system.

motor noise, which is picked up and conducted over long power lines and into sensitive equipment [Arc04]. While the US military encountered interference problems prior to World War I when trying to equip a car with a radio, little is known about the first efforts to minimize or at least find ways to counter or shield this interference. After the First World War, the radio technology which was used and perfected during the conflict, blossomed into civilian broadcasting. As more and more radio transmitters were built, it became necessary to assign different frequencies to various types of radio uses on an international basis in order to avoid interference. Aside from the harmonics generated by neighboring radio transmitters, radio receivers were in turn susceptible to adjacent channel interferences as well as to the unsuppressed impulsive noise generated by electric motors and ignition systems [Mar88]. The radio receiver which fitted into a car was consequently very susceptible to these disturbances, and it is therefore no wonder that its EMC performance has been observed and studied intensively from then on. The first IRE paper written on this subject stems from 1932, and describes the "electrical interference in motor car receivers" [Cur32] [3].

The awareness of EMC kept growing throughout the second World War, where new systems like radar emerged, and on-board radio communication became more prevalent in cars, aircraft and ships. The metal superstructure of aircraft provided excellent (yet unpredictable) energy transfer paths between various systems, and the armament and fuel tanks were consequently susceptible to ignition by sparks caused by large radio frequency fields. Additionally, intermodulation products originating from radio transmitters caused by the nonlinear electrical properties of metal joints corroded by seawater, disrupted various radio receivers on ships [Mar88]. Opposed to this, radio and radar jamming which are a form of intentional electromagnetic interference, were extensively developed and used during the second World War as well, e.g. to perturb the reception of the BBC throughout Nazi occupied Europe and to disturb the radio communication of enemy planes [Alb81].

## 3 Electromagnetic compatibility of integrated circuits

During the post-war period and through the 1960s, EMC was primarily a concern for the military, for example to control and regulate radar emissions which could (and in some cases did) cause inadvertent weapons releases by interfering with electronic fire mechanisms. Mankind will probably never know for sure how close such EMC incompatibilities have brought it to the brink of World War III, although its potentially disastrous effects can be duly appreci-

[3] The IRE, "Institute of Radio Engineers", merged in 1963 with the AIEE, the "American Institute of Electrical Engineers", to the present day "Institute of Electrical and Electronics Engineers", better known as IEEE.

ated in [Arm07] by means of real-life events [4]. The first research on the effect of electromagnetic interferences on integrated circuits (IC's) started in 1965 at the Special Weapons Center, based at Kirtland, New Mexico, USA. It is not surprising that 3 years after the Cuban missile crisis and in the middle of the cold war, the first EMC studies related to IC's investigated the effects of electromagnetic fields triggered by nuclear explosions, on electronic devices used in missile launch sites [Sic07a]. The electromagnetic pulse (better known as EMP) radiates from a nuclear detonation, and has an immense field strength. A nuclear detonation at an altitude in excess of 40 km, may disrupt and irrevocably damage electric and electronic systems up to 5000 km from the site of detonation [Kei87].

The military dominion in EMC related matters changed radically after the massive home computer proliferation starting from the 1970s. Indeed, interference problems from computing devices became a significant problem to radio and television broadcasting. In order to limit and control these interferences, various national instances as the FCC in the USA (Federal Communications Commission), or the CISPR (Comité International Spécial des Perturbations Radioélectriques) of the IEC (International Electrotechnical Commission) in Europe, started to compile a set of rules to regulate the amount of emissions, and how the measurements of these electromagnetic emissions were to be performed. In 1979, a complete issue of the *IEEE Transactions on Electromagnetic Compatibility* (vol. 21, no. 4) was devoted to electromagnetic interference problems in integrated circuits. Amongst others, this issue contained a paper describing a methodology to simulate radio frequency interferences (RFI) effects in the 741 operational amplifier designed with bipolar transistors, using computer aided simulations and macromodels [Tro79].

EMC problems and consecutive emission and immunity requirements persisted throughout the 1980s, and became even more stringent in the 1990s, owing to the growing use of electronic equipment and the ever increasing integration of different systems in the same product as well as in the same environment, invariably linking the EMC problem to the coexistence issue between circuits and systems. When different circuits and systems are densely integrated in the same appliance, the parasitic electromagnetic coupling between these circuits sharing the same printed circuit board (PCB), power supply and ground lines, is indeed a critical design parameter that can no longer be safely excluded from a product design flow. As an example, Bluetooth, GSM and WiFi services have to coexist and operate simultaneously within the close con-

[4] [Arm07] enumerates and describes the first 500 'banana skins' that have been published to date in *The EMC Journal*, published by Nutwood UK. These banana skins offer an account of real life EMC experiences, and were compiled from research reports, official documents and personal anecdotes, and vary from amusing to highly tragic situations.

finement of a modern mobile phone. Furthermore, the use of higher frequencies all contribute to an increased high frequency interference. All this means that the EMC history is repeating itself since small PCB tracks and wires pick up the disturbing signals just as easy as long power supply lines which picked up the motor noise and injected this into sensitive electronic appliances ... more than half a century ago. Currently, the victims of these electromagnetic disturbances are the integrated circuits which nowadays heavily populate and form the beating heart of almost any given electronic appliance.

According to [Deu03a], what is required to design EMC robust IC's, is amongst others: a better knowledge of how fast switching transients generate and affect electromagnetic emission, better package simulation models, better tools for simulating the EMC management on-chip, reduction of signal integrity, higher IC immunity to EMI, better control over radiated emissions, better packages with smaller parasitic elements, a controlled slew rate to reduce the $di/dt$ noise and last but not least more design engineers who understand the generation of IC's emissions and how to improve their immunity. As advocated wisely in [Deu03a], if all these points are followed, we can expect to keep up with Moore's Law for a long time, where EMC is concerned.

## 4 Scope of this book

The scope of this book is to describe the design of analog integrated circuits which achieve a higher degree of immunity against electromagnetic interference (EMI). This is in itself a very vast subject, and many research is still required in this domain, since any circuit can (and will) exhibit EMC related problems, as long as the injected EMC level is sufficiently important. The performed research has been concentrated on performing a generalized study on how IC's are affected by EMI, and what steps based on circuit modifications can be taken in order to resolve appearing EMC problems. This does of course not mean that a sloppy and an EMC-unaware PCB layout is either acceptable or justified. On the contrary, this research is meant to be an addition in increasing the global immunity of a whole system, starting from the cabling and the wire harness, to the PCB, using proper shielding where necessary and finally by improving the immunity of IC's which are connected and interconnected to this PCB in question. This work describes the studies that have been pursued, and the results that have been obtained in this matter. To this end, this book is organized as follows:

- **Chapter 1** introduces EMC from a historical perspective, and describes the scope of this work.
- **Chapter 2** starts by explaining and defining common EMC related terms. Following this, the gap which seemingly exists between electromagnetism and EMC at integrated circuit level is bridged by explaining how electro-

magnetic waves interfere and tie-in with an arbitrary IC. Unfortunately, many EMC related reference material still contains whiffs from the past, when EMC specialists used a set of rules known to themselves and which were largely based on experience and rule of thumb guidelines to solve particular EMC issues. It is the purpose of this chapter to demystify these definitions, and to illustrate briefly where these general design guidelines come from and how they should be interpreted. Finally, the standardized electromagnetic immunity measurement methods at IC level are summarized and explained.

- **Chapter 3** covers some general aspects concerning the effect of EMI in IC's. As will be shown, these appearing EMC effects can be categorized in two classes, namely a parasitic coupling, and a parasitic mixing with in particular a self-mixing resulting in a DC component, which may lead to a shift in the DC operating point (DC shift). These principles are demonstrated and derived using four design cases. The concept of DC shift is derived and illustrated in the first three case studies, discussing respectively a NMOS diode connected transistor, a source follower and a CMOS current mirror. A basic electrostatic discharge (ESD) protective structure is described in the fourth case study in order to highlight which kind of EMC interferences can be expected in them.

- **Chapter 4** discusses the effect of EMI on analog output stages. Many different output stages can be distinguished, and for this reason this chapter starts with an elementary generalization, where outputs are classified according to whether the output is driven by a transistor source, whether by a transistor drain. This basic differentiation helps to understand and solve appearing EMI problems in analog output stages from a conceptual point of view. The theoretical deductions are further elaborated and applied in two case studies, namely in an EMI resisting DC current regulator which is resistively trimmed, and in a local interconnect network (LIN) driver, which must present a high degree of electromagnetic immunity at its output. Both structures exhibit a high susceptibility to EMI in their original form, which is analyzed mathematically using the observations made in the first part of this chapter. Proposed solutions circumvent the appearing EMC problems by modifying the circuit topologies. Measurements of respective test-IC's illustrating their improved EMI performance are presented, and corroborate the general theoretical framework as well as the individual concepts behind both circuits.

- **Chapter 5** reports the effect of EMI on analog input stages. Two distinct input circuits are distinguished in this chapter. Firstly, the electromagnetic susceptibility of operational amplifiers is studied when EMI is injected into their input terminals. Existing differential pair circuits increasing the immunity to electromagnetic interference are repertoried and compared in terms of EMI induced offset suppression, current consumption and noise performance. Finally, a source-buffered differential pair structure exhibiting a high resistance to EMI is introduced. The measurements of a test-IC are described in detail, and are shown to correlate with the analytical developments. Secondly, the input stage of an instrumentation amplifier which has to resist a very high interfering common-mode voltage at its inputs is studied. As illustrated, existing input structures are very susceptible to mismatch, which translates a portion of the common-mode EMI into a detrimental differential-mode EMI component. A new input structure using current modulation is introduced and described. Simulations illustrate the superior performance and the smaller dependence on matching of the proposed input structure compared to classic structures.

- **Chapter 6** considers the effect of EMI which is conveyed through the power supply lines into linear voltage regulators. The latter are extensively used in order to regulate internal supply voltages and must, consequently, present a high degree of immunity against EMI which is injected into the external power supply terminals. First of all, the effect of EMI injected in a Kuijk-type bandgap voltage reference is studied analytically. It is illustrated how EMI couples from the supply to the reference node, and this analysis is used in order to design two EMI resisting Kuijk-type bandgap references. The measurements of a test IC comparing the original Kuijk bandgap structure to the EMI resisting ones are presented and compared to the mathematical analysis. In the second part of this chapter, the EMI susceptibility of LDO voltage regulators is studied from a conceptual point of view. It is illustrated with simulations how, applying the same design rules derived while improving the EMI resistance of a Kuijk bandgap in the first part of this chapter, the theoretical EMI immunity of LDO voltage regulators can be increased.

- Finally, the most important conclusions of this work are summarized in **Chapter 7**, and future possible research paths based on this work are briefly enumerated.

# Chapter 2

# Basic EMC Concepts at IC Level

*The FCC's Kansas City office received a complaint that the Search and Research Satellite Aided Tracking (SARSAT) system was experiencing interference from an unknown source. SARSAT is used by search-and-rescue teams to locate the radio beacon transmitters of crashed aircraft and distressed ships. Using mobile direction-finding gear, the FCC tracked the interference to a (presumably malfunctioning!) video display unit at a Wendy's restaurant.*

—Quoted from [Arm07]

## 1 Introduction

Related to the design of practical electric and electronic appliances on one hand, and to the general electromagnetic principles and theory on the other hand, EMC is an interdisciplinary scientific domain that has introduced and maintained its own typical vocabulary, conventions, definitions and design guidelines over the years. As stipulated in the previous chapter, the major focus in this work lies on the design of analog integrated circuits exhibiting a high degree of immunity against electromagnetic interferences. This chapter therefore concentrates on the general EMC issues which appear at IC level.

Standardized measurement methods were developed in order to simulate as well as replicate in measurements the appearing EMC incompatibilities in integrated circuits. Using these measurement methods to evaluate the EMC behavior of IC's as such, does not require an in-depth knowledge of EMC or electromagnetism. Quite in the same way, numerous EMC-friendly design guidelines describe what should be done in order to eliminate or at least alleviate EMC problems in electronic circuits (although the vast majority of these guidelines are solely addressing the PCB level design). One may wonder if these design guidelines can not be used as such, without any theoretical EMC knowledge.

J.-M. Redouté, M. Steyaert, *EMC of Analog Integrated Circuits*,
Analog Circuits and Signal Processing,
DOI 10.1007/978-90-481-3230-0_2, © Springer Science+Business Media B.V. 2010

The answer to this question is of course a matter of opinion: however, the bottom line is that using established measurement methods and corresponding design guidelines without any notion of where they're coming from or what restrictions they intrinsically contain, proves very often to be unfruitful and thought-constricting. Especially the latter is very undesirable since it impairs the flexibility and creativity which is required when designing electronic circuits. Electromagnetism is a scientific discipline which is unfortunately still commonly considered to be a standalone subject, dealing with antennas, transmission lines and radio waves, and therefore not directly tied to electricity and electronics. However, its impact on EMC is fundamental and profound, and its basic laws lie at the bottom of so-called rule of thumb EMC-friendly design guidelines as well as of the established and standardized measurement methods [Car95]. It is for this reason important to devote some attention to the links which exist between electromagnetism and EMC at IC level. Of course, this subject is in itself much too elaborate to be covered in full in this work. For this reason, the most basic concepts and tie-ins are discussed and presented in this chapter, offering a glimpse of what lies beyond the common rules of thumb and accepted measurement methods.

This chapter starts with a general classification of EMC terminology, and describes some frequent and palpable sources of electromagnetic disturbances. Next, a section is devoted to the link existing between electromagnetism and EMC-friendly integrated circuit design. Afterwards, the EMC issues in IC's are briefly discussed, and the main differences between digital and analog circuits are covered from a conceptual point of view where EMC is concerned. Finally, existing measurement methods for simulating and testing the electromagnetic susceptibility of integrated circuits are shortly reviewed.

## 2 Definition of EMC, EMI, EMS and EME

Many definitions are applicable in order to describe the principle of electromagnetic compatibility (EMC). The definition rendered here is the one offered in [Kei87], as it stands out because of its clearness and its unambiguity:

*Electrical and electronic devices are said to be electromagnetically compatible when the electrical noise generated by each does not interfere with the normal performance of any of the others. Electromagnetic compatibility is that happy situation in which systems work as intended, both within themselves and within their environment.*

When there is no EMC, this is due to electromagnetic interference (EMI). Quoted from [Kei87]:

*EMI is said to exist when undesirable voltages or currents are present to influence adversely the performance of a device. These voltages or currents may reach the victim device by conduction or by electromagnetic field radiation.*

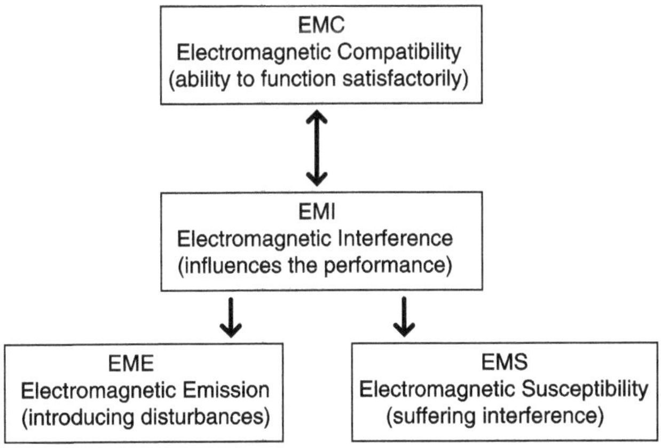


*Figure 2.1.* The used EMC terms in this work and their interrelationships, as represented in [Goe01].

This last precision is not superfluous, and a clear distinction between these two interference types must be made. To be precise, the term "radiated interference" in the above definition comprises two phenomena, namely near field coupling and far field radiation. This distinction is important and not a purely academic categorization, as will become apparent in Sect. 4.

When there is EMI, there is at least one EMI source causing an intolerable emission (be it conducted, near field coupled or far field radiated), and possibly one or more EMI victim(s) which for one or more reasons is (are) susceptible to the emanated disturbance. Electromagnetic emission (EME) is described by the International Electrotechnical Commission (IEC) as [IEV]:

*The phenomenon by which electromagnetic energy emanates from a source.*

In the same way, the IEC describes electromagnetic susceptibility (EMS) as [IEV]:

*The inability of a device, circuit or system to perform without degradation in the presence of an electromagnetic disturbance.*

Susceptibility is complementary to immunity, the latter describing to what extent EMI may be injected into a system before failures start to occur. Because the acronym for electromagnetic immunity would conflict with the one used for electromagnetic interference, this term is not abbreviated in this work: when used in the text, immunity always signifies the opposite of susceptibility. Care must be taken when using the concepts of immunity and susceptibility without distinction, since this easily leads to confusion. These four different phenomena and the way they are related to each other are represented schematically in Fig. 2.1, as in [Goe01].

## 3 Sources of electromagnetic interference

Nature contributes to electromagnetic pollution primarily by atmospheric noise (which is amongst others produced by lighting during thunderstorms) and cosmic noise [Wes01]). Lightning induces electromagnetic emissions which propagate over distances ranging up to several thousand kilometers, causing spikes or sharp random pulses in the electromagnetic spectrum. The spectral components of lightning span a wide range of frequencies, from a few Hertz to well over 100 MHz [Kei87]. Cosmic noise is a composite of noise sources comprised of:

- **Cosmic microwave background radiation:** discovered by Arno Penzias and Robert Wilson in 1965, the cosmic microwave background radiation confirms the Big Bang theory which has been predicted by George Gamow in cosmology, and it constitutes the radio remnant of the origin of our universe [Pen68]. The background radiation is isotropic, and it has a thermal black body spectrum at a temperature of 2.725 Kelvin. Its spectrum peaks at a frequency of 160.2 GHz [1] [Liv92].
- **Solar radio noise:** is proportional to solar activity and the generation of solar prominences and flares. Satellite observations have demonstrated that X-ray and ultraviolet emissions are especially intense in the heart of solar flares [Cha98].
- **Galactic noise:** with similar characteristics as thermal noise, it seems to come most strongly from the Sagittarius constellation [Kei87]. This complex radio source at the center of our Galaxy is identified as Sagittarius A, and it could equally be a plausible location for a supermassive black hole which astrophysicists believe is at the center of our galaxy [Cha98].

Several other natural noise sources and their corresponding emission spectra are enumerated in [Wes01]. Unsurprisingly, most pollution – be it environmental or electromagnetic – is man-made. Engine ignition in automotive devices, AC high-voltage power lines, microwave ovens, electric motors, communication transmitters, ... all these appliances, applications and systems contribute to an electromagnetically polluted radio spectrum [Mur03, Pat05]. These electromagnetic disturbances span a very broad frequency range, ranging from a few tens of Hz (typically 50–60, depending on the frequency of the power grid) to tens of GHz (frequency bands of modern communication systems). Extensive listings of man-made electromagnetic noise sources, intentional as well as

[1] Remarkably, the cosmic microwave background radiation contains more energy than has ever been emitted by all the stars and the galaxies that have ever existed in the history of the universe: the reason for this is that stars and galaxies (though very intense sources of radiation) occupy only a small fraction of space. When their energy is averaged out over the volume of the entire cosmos, it falls short of the energy in the microwave background by at least a factor of 10 [Cha98].

unintentional, functional as well as nonfunctional, are reported in [Wes01] and in [Kei87]. The threat associated to the criminal and covert use of intentional EMI has been discussed and illustrated with some examples and 'banana skins' in respectively [Wik00] and [Arm07].

## 4 Electromagnetism versus integrated circuit design

It is useful at this point to make a symbolic link between the elegant and complex theory of electromagnetism on one hand, and the intricate as well as exciting discipline of analog integrated circuit design on the other hand. Without doing so, the sense behind the accepted EMC measurement methods as well as widely recognized so-called EMC rule of thumb design rules is quickly lost, as has been motivated at the beginning of this chapter. A basic understanding of how both worlds tie into each other is quasi-mandatory. This is however not possible to accomplish without refreshing fundamental electromagnetic concepts, necessitating a vast array of calculations [Car95, Ida04, Sch02, Hea95]. In order to fit the present material on a few pages, the results of the computations are referred to but their derivation is omitted from the text: these exact analytical derivations can, however, be looked up in detail in the cited reference works.

Simply stated, all equipment which is using electricity or electromagnetic waves in its operation is fundamentally governed by physical laws which are elegantly merged and expressed in Maxwell's equations. In order to design and to understand the working of such equipment, Maxwell's equations or simplifications thereof (e.g. Ohm's law) are used, but only for the desired operation of the device. Indeed, owing to the huge amount of required calculations, it is usually not reasonable to examine all the possible electromagnetic interactions and couplings which are taking place in an arbitrary practical piece of equipment at the same time [Mar88]. Therefore, when considering and improving the EMC behavior of an electronic circuit, a set of design guidelines based on Maxwell's equations which minimize the likelihood of incompatibility occurrences must be used. The question of how these guidelines relate to the EMI frequencies is answered in the next paragraph.

### 4.1 Electrical length

An important step in understanding how electromagnetic waves influence a circuit's behavior is to introduce the *electrical length*, defined as the ratio of the physical length of a conductor, antenna, PCB track or device to the wavelength of the electromagnetic signal in question:

$$Electrical\ length = \frac{L}{\lambda} \tag{2.1}$$

*Table 2.1.* Electrical length of circuit components and basic physical connections for boundary EMI frequencies.

| | *Physical length* | *EMI frequency* | *EMI wavelength* | *Electrical length* |
|---|---|---|---|---|
| IC tracks | 10 μm–1 mm | 150 kHz | 2 km | 0 |
| | | 1 GHz | 30 cm | 0–0.003 |
| IC bondwires, package leads, pins | 1 mm–1 cm | 150 kHz | 2 km | 0 |
| | | 1 GHz | 30 cm | 0.003–0.03 |
| PCB tracks | 1 cm–10 cm | 150 kHz | 2 km | 0 |
| | | 1 GHz | 30 cm | 0.03–0.3 |
| External wiring | 10 cm–10 m | 150 kHz | 2 km | 0–0.005 |
| | | 1 GHz | 30 cm | 0.3–30 |

where $L$ represents the length of the conductor, and $\lambda$ stands for the wavelength of the electromagnetic signal. In general, any electric or electronic device whose electrical length is less than 1/20 or even 1/50 (in case of large impedance mismatches) can be considered as electrically short. Electrically short circuits can – depending on the desired accuracy – be fully described by basic circuit theory without having to worry about electromagnetism. On the other hand, the opposite is true for electrically long circuits: these require knowledge of electromagnetic theory in order to be solved and understood correctly [Sch02]. The major advantage of using the unitless electrical length relation resides in the fact that antennas and other radiating systems and coupling mechanisms become more easy to understand. Since the power of an antenna is proportional to its electrical length, a 50 Hz antenna, a 100 MHz antenna and a 1 GHz antenna all have the same radiation pattern and radiate with the same amount of energy if they have the same geometry with equal dimensions as measured by the electrical length, as well as identical material properties, as reported in [Sch02]. This property is rooted in the fact that Maxwell's equations are linear.

It is interesting at this point to observe how IC's, bondwires, package leads, pins, PCB tracks and external wires all relate to the electrical length, in order to check whether they are "electrically short", or "electrically long". As explained further in Sect. 8, EMC at IC level is currently measured between 150 kHz and 1 GHz. An overview enumerating the electrical length for both boundary EMI frequencies for typical circuit components and basic physical connections is presented in Table 2.1. Observe that taking the current EMC regulations into account, IC's themselves are electrically short. However, the interconnects lie very close to each other, and since the electric and magnetic field component are inversely proportional to the square of the distance in the near field as explained later on, parasitic coupling (crosstalk) may not be disregarded. Bondwires, package leads and pins start to behave as electrically long at larger EMI frequencies [Sic07b]. PCB tracks and external wiring, must be considered as being electrically long. The relevance of this will become more clear in a few paragraphs. Finally, it should be observed that the upper EMI frequency limit of 1 GHz is expected to be increased in the nearby future, meaning that IC tracks themselves will need to be considered as electrically long [Sic07c].

## 4.2 Near field versus far field

Although everybody is aware of the phenomenon of electromagnetic radiation, many misconceptions exist regarding this subject. This is mainly due to the confusing terminology as well as the fact that anything which is transmitted wirelessly using electromagnetic signals is commonly referred to as radiation. All this leads people to make basically inconsistent remarks like "disturbances owing to a 50 Hz radiation". As is explained in this section, far field radiation at 50 Hz is never encountered on Earth [Sch02].

Electromagnetic fields are basically divided into two types: near fields (storage fields) and far fields (radiating fields). Both are found mathematically when the magnetic and electric fields of an arbitrary moving point charge are developed using the Liénard-Wiechert potentials (directly originating from Maxwell's equations) [Hea95]. Extensive calculations yield that the electric ($\mathbf{E}$) and the magnetic ($\mathbf{B}$) field may be parsed into a velocity ($\mathbf{E}_v$, $\mathbf{B}_v$) and an acceleration ($\mathbf{E}_a$, $\mathbf{B}_a$) component, and that they are proportional to the distance ($r$) as follows [Hea95]:

$$\begin{aligned} \mathbf{E}_v, \mathbf{B}_v &\propto \frac{1}{r^2} \\ \mathbf{E}_a, \mathbf{B}_a &\propto \frac{1}{r} \end{aligned} \tag{2.2}$$

Integrating the Poynting vectors over the area of the sphere with radius $r$ surrounding the moving point charge, yields the results that the energy which is associated to a static, or a constant velocity field remains attached to the charge, while the interplay of magnetic and electric acceleration fields constitute radiation, which detaches itself from the charge and travels off to infinity as an independent electromagnetic system. Both fields are now briefly clarified [Sch02, Hea95, Car95].

- **The velocity field** is commonly referred to as near field, reactive field or storage field, because it stores and transports energy in the near area of its source. Storage fields equally disappear when their source is turned off. Consider as an example an ideal inductor $L_1$ which is driven by an AC source. Ideally, no energy is lost in this inductor, since it generates a storage field, pumping power into this field which at the same time returns power to the circuit. This energy cycling is responsible for the 90 degrees phase shift between the voltage and the current. However, the moment a second inductor $L_2$ is placed in the near vicinity of the first one, the field from $L_1$ couples into $L_2$. If $L_2$ is shunted with a load resistor $R_L$, current flows through this resistor, and the reactive field allows energy to be transferred from the AC source to the resistor. Not surprisingly, this circuit behaves like a transformer, whereby energy is sapped from the AC source driving $L_1$ and dissipated in load resistor $R_L$. Note that the same effect could be achieved using ideal capacitors. A reactive field can therefore store energy, transport energy, or do both at the same time [Sch02].

- **The acceleration field** is commonly referred to as far field, or radiating field, because it radiates energy. As stated previously, these radiating fields propagate forever, even after their source is turned off. Antennas are focused on launching those fields, so that they propagate from the source, regardless of a receiving antenna [Sch02]. This energy loss appears as an energy dissipation across a resistor which is connected to the source: this resistance is called the radiation resistance. Observe that since acceleration can be positive or negative, energy is equally radiated upon deceleration. In the particular case when electrons are projected into a material in which they are stopped or slowed down, radiation results (as with X-ray beams) [Sch02].

The boundary between the near field and the far field is generally considered to lie around $2D^2/\lambda$, where $D$ is the size of the transmitting antenna [Ben06]. For a dipole antenna, the reactive field becomes typically negligible at distances varying from $3\lambda$ to $10\lambda$ of the dipole [Sch02]. This explains why ordinary optical sources (e.g. a light bulb) appear as radiating sources and not as reactive sources, unless they are approached closer than a few μm. Referring to the beginning of this section, it is equally clear that a "disturbance owing to a 50 Hz radiation" is not possible on Earth because the disturbed appliance should be situated at a distance of more than thousand kilometers from the source to even get to the edge of the near field. However, disturbances associated to the 50 Hz power lines are possible (and in fact, occur quite often) as a result of a near field coupling [Sch02]. This differentiation is important when studying and deriving the basic concepts to reduce these disturbances [2].

[2] This explanation cultivates and sustains the impression that radiation is not present in the near field, while coupling does not occur in the far field. This practical approach is not mathematically complete, as is explained here. The electromagnetic field which is emanated by a transmitting antenna is expressed by the Poynting vector $\mathbf{E} \times \mathbf{H}$, in which $\mathbf{E}$ and $\mathbf{H}$ are the electric and magnetic field. Since an antenna is modeled

## 4.3 Radiation of a conductor

Recall from the previous paragraph that only accelerated charges radiate in the far field. In the near–constant velocity–field, electromagnetic energy is stored, transported and coupled. When a charged particle moves in a circle, or in any oscillatory manner, it experiences a sinusoidal acceleration. When a constant DC voltage $V$ is connected across a conductor with resistance $R$, the current through the wire is defined by Ohms law ($I = V/R$). Although the net electron flow in the conductor is traveling at a constant speed, individually, the electron movement is quite random, and multiple collisions happen in-between the electrons, causing heat radiation [3]. The larger the current, the more collisions, and the more the conductor dissipates and radiates heat. Some of this heat radiation is propagated at lower energies, in the microwave and radio bands: this is the troublesome thermal white noise that is impeding the design of low noise circuits and low noise IC's. But, except for the heat, there is no radiation because the net current flow is constant [Sch02].

Assume now that the voltage source slowly oscillates in time. As long as the wavelength associated to the oscillating frequency is much larger than the length of the conductor, the electrical length is very small as expressed in (2.1), meaning that the acceleration (and the corresponding radiation) is small. This explains why electrically short antennas are not very efficient radiators [Sch02]. Mathematically, the radiated power ($P_{rad}$) of a conductor is found by calculating the time-averaged power density in the far field, and surrounding it by a sphere of radius $r$ to calculate the total power traversing the outer surface of the sphere [Ida04]. As an example, for an electrically short Hertzian dipole antenna, the result of this calculation yields:

$$P_{rad} = I^2 \frac{\eta \cdot \pi \cdot (l')^2}{3 \cdot \lambda^2} \tag{2.3}$$

where $l'$ is the length of the Hertzian dipole antenna, $\lambda$ the wavelength and $\eta$ represents the *far field wave impedance* of an electromagnetic wave, which is the ratio of the transverse components of the electric and magnetic fields in the far field. This ratio is equal to $\sqrt{\mu/\epsilon} \approx 377\ \Omega \approx 120 \cdot \pi\ \Omega$ in a lossless

by a RLC circuit, the generated electromagnetic field can be subdivided into a real field (generated by the resistive component) and a reactive field (generated by the capacitive and inductive components). The latter is predominant in the close proximity of the antenna, causing electromagnetic coupling: although the reactive field decays exponentially with increasing radius from the antenna, it is nevertheless present in the far field. Conversely, the real field, causing radiation, is expressed by the real part of the Poynting vector: it is dominant in the far field, but equally present in the near field. However, the near field is dominated by the reactive field (responsible for the electromagnetic coupling), while the opposite is true in the far field, which is pervaded by the real field (responsible for radiation). Strictly speaking, therefore, radiation at 50 Hz *is* encountered on earth: however, it can largely be neglected since its near field coupling largely dominates.

[3] Corresponding to radiation in the infrared region.

medium [4]. The far field wave impedance is a constant, which describes the physical transmission properties of a homogeneous medium. Observe that the radiation power is proportional to the current squared and the antenna length squared. It also depends on the intrinsic impedance of the medium in which the antenna radiates, and it is inversely proportional to the square of the wavelength [Ida04]. Using the definition of electrical length in (2.1), the radiated power is therefore proportional to:

$$P_{rad} \propto I^2, \quad (electrical\ length)^2 \tag{2.4}$$

Not surprisingly, previous expression indicates that high frequency signals radiate more readily than lower frequency ones over the same PCB track or conductor.

When the voltage source oscillation is increased even more, the AC current causes destructive interferences in the conductor: the radiation power is from then on no longer directly proportional to the antenna length squared, but follows more complex patterns [Sch02]. The associated integrals are quite difficult to resolve, but can be integrated numerically using integration methods. A full account of these calculations is provided in [Ida04]. The same calculations also explain why the well-known half wave dipole antenna is made slightly shorter than half a wave in practice [5] [Set97].

## 4.4 Basic EMC antenna concepts

Antenna theory and design is a very complex and elaborated research field, as can be duly appreciated in the quote at the beginning of this chapter. The object of this paragraph is to identify how unwanted and parasitic antennas appear in practical PCB and IC design. In this context, the two most basic types are studied, namely an electric dipole (Hertzian) antenna and a magnetic dipole (loop) antenna. As observed further on, the derived principles can be applied to reduce unwanted EMI coupling and pick-up from near and far fields. An electric dipole antenna and a magnetic loop antenna are depicted respectively in Figs. 2.2a and 2.2b. Observe that the magnetic field is not shown in the electric dipole antenna representation, and that the electric field is not drawn on the magnetic dipole antenna illustration: they have simply been omitted in order to not overload the figure unnecessarily. Furthermore, in the near field, the coupling is electrical for an electric dipole antenna, while solely magnetic for a magnetic dipole antenna as will be shown later on.

---

[4] The *far field wave impedance* is often defined as the wave impedance in the specialized literature. The reason why it is identified as "far field wave impedance" in this work, stems from the fact that a "near field wave impedance" is commonly distinguished as well: confusingly, both wave impedances describe different physical properties while being very often identified by the same term, which easily leads to confusion.

[5] In the strict sense, a half wave dipole antenna has a radiation impedance of $(73 + j40)\ \Omega$: making the antenna slightly shorter cancels out the reactive component.

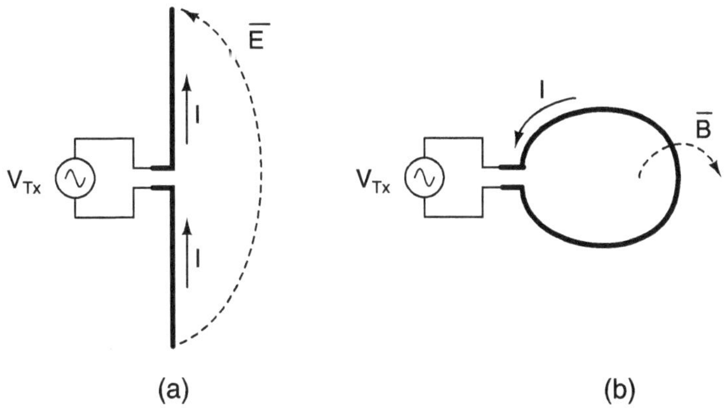


*Figure 2.2.* (a) Electric dipole (Hertzian) antenna. (b) Magnetic dipole (loop) antenna.

In the far field, radiation occurs for both antennas. In the first case, the electric dipole antenna generates an electric field. Consequently, this electric field produces a magnetic field, and the two fields propagate from the antenna. Similarly, in the second case, the loop antenna produces a magnetic field around the conductor: this magnetic field generates an electric field, and the two propagate away from the antenna. Upon arrival at the receiver, currents are induced on the receiving antenna: if the receiving antenna is an electric dipole antenna, it will receive the electric field, and the opposite is true for a loop antenna, which will receive the magnetic field mostly [Set97]. An ideal electric and magnetic dipole antenna with the same dimensions have the same radiation impedance and radiate the same power. Electric and magnetic components of far field waves are fixed by their far field wave impedance [Sch02]. The far field wave impedance is equal to the ratio of the magnitudes of the transversal electric and magnetic fields components, as defined anteriorly, and is approximately equal to 377 Ω in free space [Hea95].

This situation changes in the near field. The electric power of a transmitting electric dipole antenna is coupled on a nearby electric dipole antenna through its electric field, in the same way as capacitor plates (Fig. 2.3a). Parallelly, a nearby magnetic dipole antenna picks up the electric power which is coupled from an emitting magnetic dipole antenna through its magnetic field (Fig. 2.3b). Taking the ratio of the electric and magnetic field components in the near field, results in a quantity which has the same dimensions as the far field wave impedance, although it is a function of the distance to the antenna, while the far field wave impedance is a constant. Confusingly, this former parameter is equally defined as the wave impedance in most of the EMC related literature. In this work, this function is identified as the *near field wave impedance*, in order to distinguish it from the far field wave impedance. The near

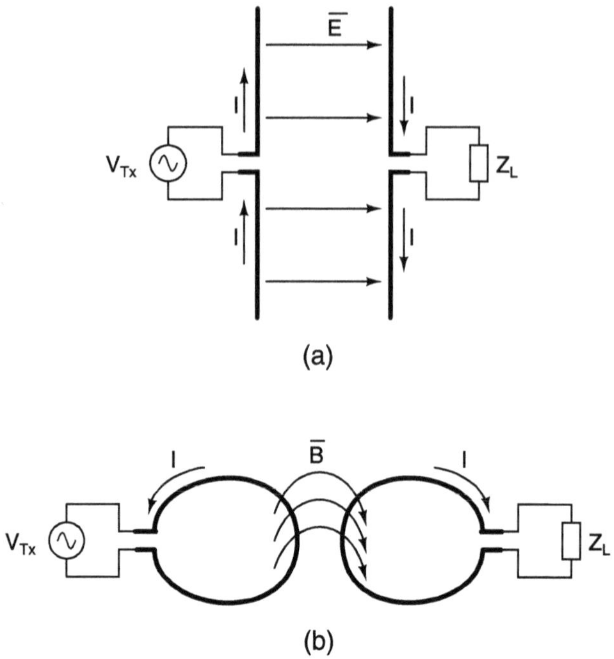


*Figure 2.3.* (a) Coupling between electric dipole (Hertzian) antennas. (b) Coupling between magnetic dipole (loop) antennas.

field wave impedance depends on the structure of the source, namely, on the antenna (or parasitic antenna type structure) which is generating the electromagnetic field. In particular, in the near field of an electric dipole antenna, the near field wave impedance ($Z_E$) increases, while in the near field of a magnetic dipole, the near field wave impedance ($Z_H$) decreases [Goe01]:

$$Z_E = \frac{|E|}{|H|} = \frac{1}{2 \cdot \pi \cdot \epsilon \cdot f \cdot r}, \qquad Z_H = \frac{|E|}{|H|} = 2 \cdot \pi \cdot \mu \cdot f \cdot r \tag{2.5}$$

where $|E|$ and $|H|$ are the electric and magnetic field components which are perpendicular to the propagation direction, $r$ represents the distance to the respective electric and magnetic dipole antenna, and $\epsilon$, $\mu$ equal the permittivity and permeability of the transmission medium, respectively. These relationships have led to the use of terms *high impedance electric field* and *low impedance magnetic field* [Goe01].

## 4.5 Radiated, induced and conducted disturbances

Throughout EMC regulations and corresponding literature, two types of disturbances are distinguished, namely conducted and radiated disturbances. In view of previous paragraphs, this is not correct, since both near field coupling as far field radiation have been lumped under the term "radiated". However, many authors prefer to talk about radiated disturbances, and make the near field and far field approximations ulteriorly. They define radiated disturbance as any interference which is transferred through a non-metallic (or simply non-conductive) medium by an electromagnetic field. Even the IEV (International Electrotechnical Vocabulary) adopted by the IEC in all their standards defines electromagnetic radiation as [IEV]:

1 *The phenomenon by which energy in the form of electromagnetic waves emanates from a source into space.*

2 *Energy transferred through space in the form of electromagnetic waves.*

Clearly, no difference is distinguished between near field coupling and far field radiation. This easily leads to confusion and misplaced observations, since there are many practical differences between near field coupling and far field radiation. As described previously, near field waves are usually dominated by either the electric either the magnetic component. As an example, the symmetrical design of a half wave loop antenna (resembling a magnetic dipole) generates a high magnetic field around the loop. On the other hand, a classic Hertzian half wave dipole antenna (resembling an electric dipole) generates a strong electric field perpendicular to the dipole. In this work, the same notations and definitions as in [Sch02] are followed, and the electromagnetic disturbances are grouped into three distinct categories:

- **Induced interferences (near field coupling):** unlike radiated interferences, near field waves are very often dominated by either the magnetic, either the electric component. Circuits pick up radiated energy if they contain electric or magnetic antenna-like elements, like dangling conductors and loops. It has previously been illustrated that long conductors are susceptible to electric fields, while large conductive loops are especially susceptible to magnetic fields. In addition, the better the receiving antenna's impedance is matched to the near field wave impedance at that particular distance from the transmitting antenna, the more energy is transported. As expressed in (2.5), electric near fields have a high near field wave impedance while magnetic near fields have a low near field wave impedance. Therefore, *high impedance circuits or nodes are particularly susceptible to electric near fields, while low impedance circuits or nodes are particularly susceptible to magnetic near fields* [Sch02]. Finally, near field interferences increase with larger fields, higher frequencies and shorter distances. Induced interferences are often referred to as capacitive and inductive crosstalk, depending on the coupling being electric or magnetic.

- **Radiated interferences (far field radiation):** these interferences are constituted of purely electromagnetic transversal waves. They will therefore easily radiate (and parallelly, be received) on long loops (forming loop antennas) and long conductors (forming Hertzian antennas). Several theoretical analyses have been developed to describe the effect of radiation on e.g. an interconnecting cable or a transmission line [Kon94].

- **Conducted interferences:** this type of interferences comprises the unintended signal energy that leaves an integrated circuit through its outside connections and propagates through a conductor, e.g. a metal wire or a PCB track. Conducted interferences are in general caused by simultaneous switching noise (SSN), which generate fluctuations in e.g. a power bus. Because of these disturbances, the signals which are referenced to a particular power bus may exhibit high frequency voltage fluctuations if there is not enough decoupling or if these interferences are particularly important. The strongest currents are usually flowing in the power supplies and the ground pins [Ben06]. Consider the example of conducted electromagnetic noise which is generated on the power lines in a switched mode power supply (SMPS) [Car94]. These conducted interferences are usually dealt with using proper filtering and a good grounding strategy. When two or more current loops share a common conductor (e.g. the ground plane), one current loop may influence and alter the second current loop: this effect is identified as crosstalk via a common impedance. These issues are usually resolved by using adapted layout techniques, like point coupling (to each loop its own conductor) or a strip shaped reference (by providing all the tracks on one side of a circuit with a strip-shaped reference), and by keeping the common impedance at a low value [Goe01].

## 4.6 Practical example

In order to illustrate the implications enumerated in the previous paragraphs from a practical point of view, consider the PCB layout which is depicted in Fig. 2.4a. An arbitrary integrated circuit on this PCB is supplied by two tracks, connecting its positive ($V_{dd}$) and negative ($V_{ss}$) supply terminals to an ideal DC voltage source $V_{DC}$ (e.g. a lithium battery). As observed in Fig. 2.4a, the PCB traces form a small square-shaped loop, with 2 cm side length. Assume that there is a time-varying, low frequency magnetic field, which is perpendicular to this loop (Fig. 2.4b). The parasitic EMI voltage generated between the supply terminals of the IC ($V_{emi}$), is expressed by the Maxwell-Faraday equation [Arc04]:

$$\oint \mathbf{E} \cdot dl = -\iint \frac{\partial \mathbf{B}}{\partial t} \cdot d\mathbf{S} \qquad (2.6)$$

where **E** and **B** represent respectively the electric and the magnetic field. As long as the wavelength of the magnetic field is large compared to the length of the loop, then the magnetic flux is constant over the loop area, and previous

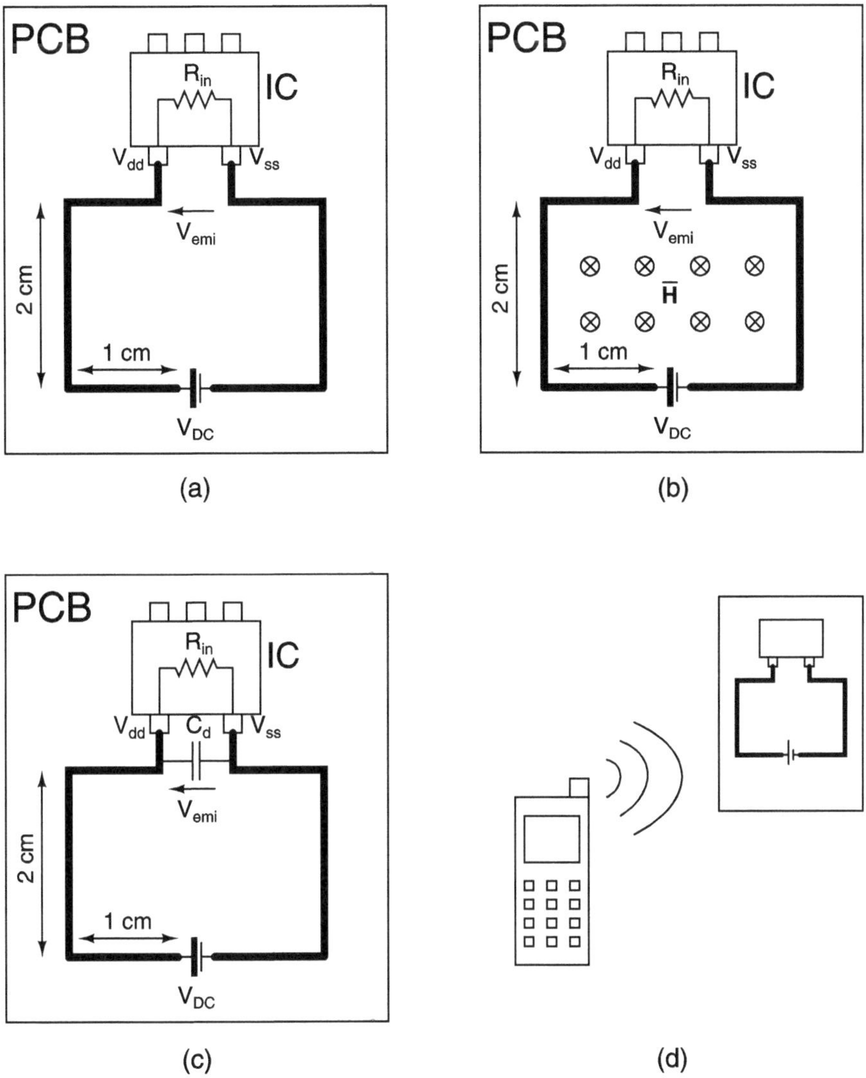


*Figure 2.4.* (a) PCB with the supply tracks connected to the IC forming a small loop. (b) With the presence of a magnetic field perpendicular to the loop. (c) With a decoupling capacitor. (d) Nearby a transmitting mobile phone.

equation can be rewritten as the Faraday's law of induction [Arc04]:

$$V_{emi} = \mathcal{E} = -\frac{d\Phi_B}{dt} = \omega \cdot \mu_o \cdot |H| \cdot A \tag{2.7}$$

where $\mathcal{E}$ is the electromotive force, $\Phi_B$ is the magnetic flux, $\omega$ is the angular frequency of the time-varying magnetic field, $\mu_o$ is the permeability of free space, $|H|$ is the magnitude of the magnetic field and $A$ is the surface enclosed by the loop. Assuming that the frequency of the time-varying magnetic field is equal to 1 MHz and that its field strength is equal to 2 A/m, the induced EMI voltage at the supply terminals of the IC is derived as follows:

$$\begin{aligned} V_{emi} &= (2 \cdot \pi \cdot 10^6 \text{ [Hz]}) \cdot (4 \cdot \pi \cdot 10^{-7} \text{ [H/m]}) \cdot (2 \text{ [A/m]}) \cdot (0.02^2 \text{ [m}^2\text{]}) \\ &= 6.3 \text{ mV (RMS)} \end{aligned} \tag{2.8}$$

Unfortunately, this value increases with the frequency of the time-varying magnetic field, and can therefore attain considerable values. As an example, if the frequency of the magnetic field is increased to 10 MHz, the induced EMI voltage is equal to 63 mV. For this reason, it is important to decouple the supply lines very close to the IC pins (and preferably, add internal decoupling capacitors as well). Consider the case when a decoupling capacitor of 100 nF is placed close to the IC terminals, as illustrated in Fig. 2.4c. Assuming that the on-chip supply impedance between $V_{dd}$ and $V_{ss}$ ($R_{in}$) is negligible compared to this 100 nF capacitor, and assuming that the series inductances of the PCB tracks are approximated using the rule of thumb predicting 1 nH per mm conductor length [Goe01], yielding a total inductance of 80 nH for the full length of the loop, the voltage shunted across the supply terminals is reduced to 2 mV instead of 63 mV. Previous example illustrates the need of minimizing the occurrence of conductive loops, and highlights the necessity of placing decouple capacitors close to the IC pins.

Consider now that a high frequency electromagnetic field (e.g. providing from a mobile phone) surrounds the PCB in question. Since the loop formed by the PCB tracks has a length which is equal to 8 cm, it behaves as a half wave loop antenna, receiving the GSM frequencies situated at 1800 and 1900 MHz:

$$\frac{\lambda}{2} \approx 8 \text{ cm} \rightarrow f \approx \frac{c}{\lambda} \approx 1.875 \text{ GHz} \tag{2.9}$$

The transmission equation developed by Friis, provides a straightforward way of calculating the power which is ideally received by an antenna ($P_r$), from another antenna some distance away, transmitting a known amount of power ($P_t$) [Set97]:

$$\frac{P_r}{P_t} = G_t \cdot G_r \cdot \left(\frac{\lambda}{4 \cdot \pi \cdot R}\right)^2 \tag{2.10}$$

where $G_t$, $G_r$ express the antenna gains of respectively the transmitting and receiving antennas, and $R$ is the distance between both antennas. Consider that a nearby mobile phone, which is situated at 1 m distance of the PCB, is transmitting at 1.875 GHz, with a peak transmitted power of 2 W, as depicted in Fig. 2.4d. Assuming out of simplicity that the mobile phone's antenna is a half wave dipole antenna, means that both antenna gains are equal to 1.64. Inputting the previous data in transmission equation (2.10), yields:

$$P_r = (2\ [\mathrm{W}]) \cdot (1.64)^2 \cdot \left(\frac{0.16\ [\mathrm{m}]}{4 \cdot \pi \cdot 1\ [\mathrm{m}]}\right)^2 = 0.87\ \mu\mathrm{W} \tag{2.11}$$

Ideally, a maximal power of 0.87 μW is received on the PCB tracks. Considering that at the frequency of interest, the internal resistance seen in the supply terminals on-chip ($R_{in}$) equals 75 Ω, and consequently matches with the loop antenna impedance, the EMI voltage between the supply pins is calculated as follows:

$$V_{emi} = \sqrt{P_r \cdot R_{in}} = \sqrt{0.87\ \mu\mathrm{W} \cdot 75\ \Omega} = 256\ \mathrm{mV\ (RMS)} \tag{2.12}$$

As is apparent from previous numerical examples, no big loops are necessary to generate significant interference levels at IC terminals. Maximal care must therefore be ensured while designing PCB and circuit layouts.

## 5 Intra-chip versus externally-coupled EMC

EMC issues associated with integrated circuits are generally classified as externally-coupled EMC or as intra-chip EMC [Ben06]:

- Externally coupled EMC problems result when noise which is generated externally interferes with the IC (EMS), or conversely, when noise generated in the IC, interferes with circuits and devices which are off-chip (EME). In this work, the former is considered, since this research is focused on the design of analog integrated circuits which have a higher degree of immunity against EMI. Owing to the small size of integrated circuits, they are in themselves not easily disturbed by radiated and induced disturbances because the on-chip interconnections they harbor are tiny and too small to function as effective antennas (refer to Table 2.1) [Mey03]. Bondwires, package leads, leadframe and pins may intercept radiated, induced and conducted high frequency disturbances. However, the main contribution comes from the noisy and relatively long PCB tracks to which IC's are ultimately connected [Fio01]. Depending on the total levels of conducted EMI which are present on such a PCB track, an unfortunate IC may not work correctly any longer: even worse, it may not work at all (Fig. 2.5). Adverse radiated and induced EMI effects can be alleviated by proper PCB layout and shielding techniques ([Goe01]), as observed previously. Preventing conducted EMI to access or emanate to or from a given IC pin, depends on the circuit to which the PCB track is connected. For instance, common-mode chokes and other discrete components (like decoupling capacitors)

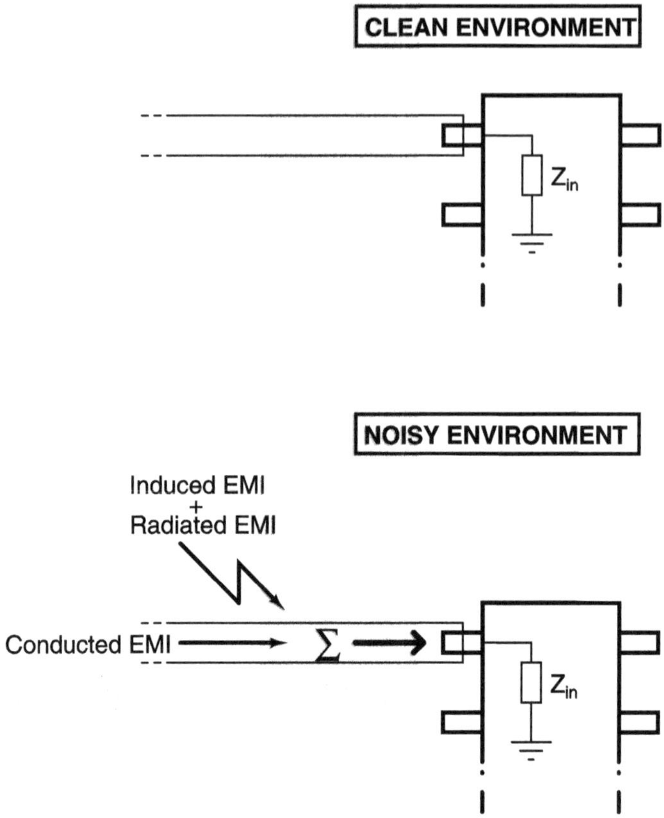


*Figure 2.5.* Schematic representation of induced, radiated and conducted EMI injected on a PCB track connected to an IC.

reduce conducted EMI on PCB tracks: however, their presence is not always wanted nor possible. Proper IC design should therefore focus on reducing the conducted EMI which is injected into PCB tracks (EME aspect). On the other hand, a certain robustness is required from integrated circuits themselves, meaning that they should be able to withstand a certain level of total conducted EMI before their correct operation is impaired. This is especially true when designing and processing IC's that are connected to unspecified PCB's, or to PCB's that are not EMC-wise characterized. Ideally, combining a proper PCB layout which reduces radiated and induced EMI, with IC's which produce less conducted EME and have a decreased EMS, leads to a full electronic system which is EMC. Conjunctively, previous sentence illustrates the necessity of including Sect. 2 and using indubitate definitions and abbreviations.

- Intra-chip EMC problems occur on the same IC, when a signal or noise created in one or more (sub)circuits interferes with the operation of another circuit block. Because on-chip distances are small, radiated intra-chip interferences are not occurring, because the far field stretches outside the IC itself. However, induced and conducted interferences are likely to occur. This results in two common IC problems, namely crosstalk and simultaneous switching noise [Ben06]:
  - **On-chip crosstalk** between two circuits or circuit elements is defined as the ratio between the unintentional signal voltage appearing across a load impedance to the signal voltage in the source circuit. Basically, three types of parasitic coupling may result in crosstalk: electric field coupling, magnetic field coupling and common impedance coupling [Goe01]. Common impedance coupling occurs when multiple current paths share the same conductor. The finite impedance of the latter generates a voltage drop which appears in the current loops sharing this conductor, and which is proportional to the total current flowing through it, as well as to its impedance. Electric field coupling can be represented by a capacitor between two tracks: in the same way, magnetic field coupling can be modeled by coupled inductors. On-chip crosstalk can be reduced by a good routing strategy [Cat95b].
  - **Simultaneous switching noise** (SSN) is a particular case of common impedance crosstalk when subcircuits on a same IC share the same power distribution bus. It is also known as ground bounce, power bounce or $di/dt$ noise. It can be reduced by using on-chip decoupling capacitors and by observing a consistent grounding strategy [Ben06]. Distortion measurement results on the output waveform of an integrated opamp owing to substrate noise generated by surrounding logical circuitry have been presented in [Cat95a].

## 6 Analog versus digital integrated circuits

Digital integrated circuits are inherently less susceptible to EMI than their analog counterparts: this stems from the fact that digital circuits have the benefit of using thresholds between logic levels, and are hereby predisposed to have a natural resistance against interferences. However, it should be pointed out that although digital integrated circuits exhibit a lower susceptibility to EMI, this does not mean that they are completely immune to it. EMI has been observed to have two distinct effects on digital devices, namely false switching (static failure) which occurs when the EMI level is large enough to change the logic state of a digital signal, and EMI induced delay, which is the change of propagation delay owing to the EMI. It has been illustrated that the latter occurs at much lower amplitudes of EMI than the former [Lau95, Cha97]. In

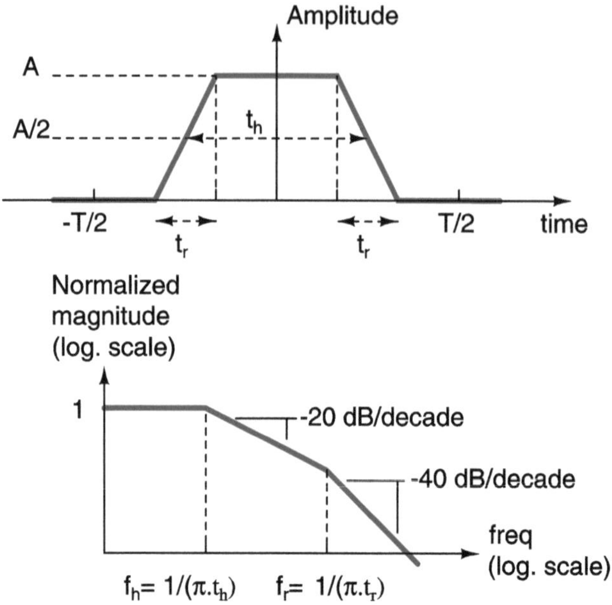


*Figure 2.6.* Shape and corresponding emission spectrum of a trapezoidal signal.

a worst case situation, depending on the total EMI level, digital integrated circuits can botch a complete data operation because some significant bits were permanently flipped into another state owing to a particularly strong EMI injection. This evidently may lead to a completely false information processing from which the system can not recover easily, and that may even require a reset [Tro85]. Conversely, the same disturbance would have barely caused a brief "crackle" in an analog circuit [Goe01]. Nevertheless, as long as realistic EMI levels are considered and if some basic precautions are taken in order to reduce and prevent the injection EMI into a particular digital integrated circuit, digital integrated circuits exhibit a higher immunity to EMI than analog ones, because of their threshold levels.

Additionally, digital circuits typically have a very fast switching behavior: this causes sharp transients, which induce a lot of high-frequency components in the electromagnetic spectrum, consequently increasing the EME. Since analog circuits process the signals in a continuous way, they tend to have a much smaller emission spectrum. As an example, Fig. 2.6 depicts the approximated spectrum of a periodic normalized trapezoidal signal, using the "three-straight-line" approximation described in [Goe01]. Observe that at frequencies above $f_h$, the asymptotic amplitude of the spectral components is inversely proportional to frequency ($-20$ dB/decade) and at frequencies above $f_r$ it is inversely proportional to the square of the frequency ($-40$ dB/decade). This underlines the necessity of reducing fast switching times, in order to achieve a smaller EME spectrum. Other design possibilities to reduce the EME in IC's include a reduction of the clock frequency, and the use of a small resistor in the power supply lines in order to dampen the oscillations generated by fast switching [Loc04]. Dedicated design techniques are used to alleviate EMI in digital IC's: as an example, logic family comparisons show that the enhanced current steering logic (ECSL) constitutes the best compromise in terms of performance and induced power supply noise [Zho08].

## 7 EMC in automotive applications

The automotive industry is particularly interested in increasing the EMC performances of electronic circuits and systems, since the automotive electromagnetic environment can be very severe and (owing to the inherent mobility of automotive applications), most unpredictable [IET]. In order to ensure that vehicle accidents are not caused as a result of EMC incompatibilities, vehicle manufacturers as well as electronic sub-assembly (ESA) companies go to enormous lengths (driven by severe product liability legislation) to ensure that their vehicles do not suffer from EMC problems. It has been claimed that occasional and untested EMI events that could cause a safety incident only once during a 10-year vehicle life, can still expose drivers to safety risks comparable with those of the world's most dangerous occupations [Arm08] [6]. In the next few years, the importance of EMC-proof applications within a vehicle is bound to increase even more, since fully electronic braking, steering and anti-collision systems are likely to be introduced in the present and nearby future. This implies that electromagnetic compatibility is of paramount importance in order to assure the correct functioning of an automobile [Ale08].

## 8 Immunity measurement methods for IC's: IEC 62132

Clearly, no equipment can sustain gracefully unlimited levels of electromagnetic aggression, without suffering an impaired or reduced operation at a certain point. When designing to achieve an increased EMC behavior, a realistic assessment of the threat levels during normal operation must be made [Mar88]. EMC measurement setups for automotive electronic systems are defined in standards such as in the International Special Committee on Radio Interference

---

[6] Quoting from [Arm08]: *"A simple analysis based on reasonable assumptions for a 6-cylinder engine at 2000 rpm with spark-ignition transients lasting 50 ns, shows that in each minute there is a 0.001% likelihood of an overlap of at least 50% with a 100 ns transient that occurs once every minute (on average, for example due to the actuation of an electric motor or solenoid). If the vehicle is driven for 1 hour/day, 5 days/week, 40 weeks/year, the likelihood of such an overlapping pulse event is 12% per year. And if the overlapping pulses caused an electronic sub-assembly (ESA) to malfunction with a 1% chance of death, the driver would have a risk of death of 0.12% per year. This compares with a death rate of about 0.1% per year for very hazardous occupations (e.g. oil industry divers)".*

(Comité International Spécial des Perturbations Radioélectriques) (CISPR) 25 for parasitic emissions, and in the International Organization for Standardization (ISO) 11452 for susceptibility to EMI [Ram09]. Since IC's are generally the main cause of EMI related malfunctioning and disturbance in electronic equipment, there has recently been considerable demand for simple, reliable, and standardized measurement methods focusing only on IC's.

The International Electrotechnical Commission (IEC) is one of the international standards organizations which are addressing the need for standardized IC EMC test methods. The IEC's IC EMC standards are sponsored by the IEC sub-committee (SC) 47A (integrated circuits), which is a part of the IEC technical committee (TC) 47 (semiconductor devices). SC 47A created working group (WG) 9 to prepare international standards for test procedures and measurement methods to evaluate the EMC of ICs [Car04]. Where possible, WG 9 coordinates the preparation of its standardized test methods with methods standardized or in progress with industry and national standards bodies including, but not limited to, the Society of Automotive Engineers (SAE) in the United States and the Verband der Elektrotechnik, Elektronik und Informationstechnik (VDE) in Germany [Car04]. SC 47A, WG 9 has released two main standards for measuring the EMC of integrated circuits: the first one (released in 2001) for measuring radiated and conducted emission [IEC 61967], and the second one (released in 2003) for measuring immunity [IEC 62132]. A short but very comprehensive overview describing previous standardized immunity measurements at IC level is given in [Car04] and in [Ben06]. Nowadays, the upper EMI frequency used in the actual EMI immunity measurement methods is limited to 1 GHz. Owing to the higher process integration, higher switching speeds and higher circuit complexity, the demands for measurements at higher EMI frequencies grow stronger, and it is very likely that this upper limit will be stretched to 3 GHz in the near future [Sic07c]. Following the requirements set by the International Technology Roadmap for Semiconductors (ITRS) for the coming years [ITRS], it is expected that even higher EMI frequencies will need to be addressed in the measurements. Future trends about IC technology and a corresponding tentative EMC roadmap until 2020, with a strong focus on embedded system-on-chips (SOC) for automotive and consumer electronics applications, is presented in [Sic07c]. Promising research results addressing the EMI frequency-band between 3 and 10 GHz results have been published: one of these is the near field scan immunity (NFSI) measurement method [Boy07]. Above 10 GHz, dedicated IC measurement methods do not exist yet [Sic07c]. Future plans include EMC measurements up to 40 GHz, but much research is still needed in this area [Sic07c].

- **TEM cell and GTEM cell:** The transverse electromagnetic mode (TEM) cell, as well as its high frequency variant – Gigahertz TEM (GTEM) cell – are used for measuring the IC immunity to electromagnetic fields [Ben06].

The TEM cell is nothing else than an expanded rectangular waveguide with an inner conductor which is called the septum. Electromagnetic Interference is injected in the septum, and a test PCB containing the IC to be measured is inserted in an aperture on the outer wall of the TEM cell, with the chip inside the cell. The maximum frequency that can be used in the TEM cell is set by the resonance of the lowest higher order mode, which is dependent on the size and the shape of the cell. Typical TEM cell dimensions can handle a 200 to 300 MHz cut-off frequency. The GTEM cell was designed to overcome the frequency limitations of the TEM cell, and so it stretches up to frequencies of several GHz (typically 18 GHz). Since the TEM cell is a radiative measurement method, it is quite cumbersome to use this method as such in circuit simulations. For many integrated applications, the TEM/GTEM cell measurements constitute the final EMC compliance tests.

- **Workbench Faraday cage (WBFC):** The workbench Faraday cage is a standard method for carrying out conducted immunity measurements [Ben06]. However, the scope of this measurement setup is very restricted, since it is only applicable to electronic products that are connected to external wiring: it is therefore not a suitable measuring method to measure e.g. small wireless appliances. Finally, it is again not practical to use this method as such in circuit simulations.

- **Bulk current injection (BCI):** The measurement reproduces the induced current that is generated in the real world by electromagnetic fields which are coupling into the wires of a system [Ben06]. Two current probes are used: one for injecting the disturbance in the wire, and the other one for measuring the level of injected current. The EMI frequency to be measured ranges from 10 kHz to 1 GHz according to the IEC standard, although in practice, the upper EMI frequency does not exceed 400 MHz [Ben06]. The main problem which prevents this measurement method to be used in circuit simulations is the fact that the magnetic coupling between the current probe and the wire is not exactly known. The typical BCI measurement configuration is represented in Fig. 2.7. For many integrated applications, the BCI measurements are used for EMC pre-compliance testing.

- **Direct power injection (DPI):** In this measurement setup, the EMI disturbance is injected into the pin of a component through a decoupling block [Ben06]. In practice and by default, this decoupling block is a capacitor ($C_c$) of 6.8 nF. The source impedance of the EMI source $R_s$ is set to 50 Ω. The default value of protection resistor $R_p$ is 0, although it may be increased up to 100 Ω if the application requires it. The EMI frequency to be measured ranges from 150 kHz to 1 GHz. The DPI measurement setup

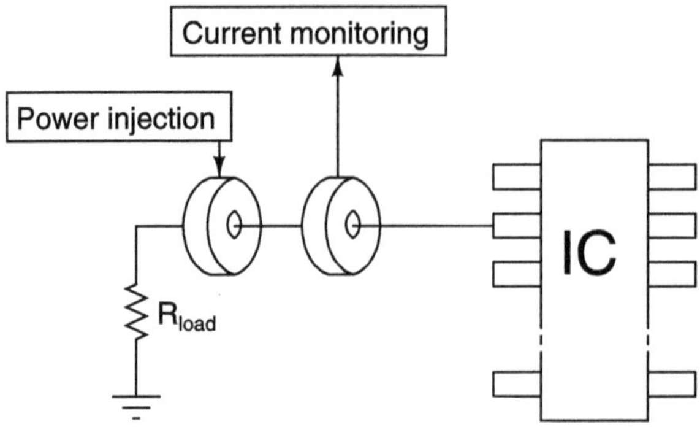


*Figure 2.7.* Bulk current injection measurement setup.

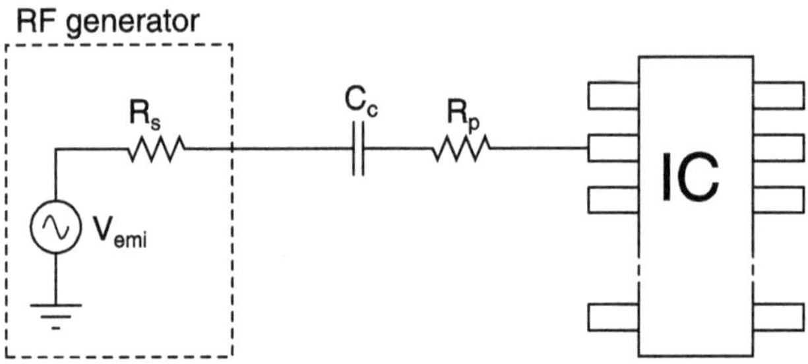


*Figure 2.8.* Direct power injection measurement setup.

*Table 2.2.* DPI injected power levels.

| *Zone* | *Forward injected EMI power (W)* | *EMI source voltage amplitude (V)* | *Notes* |
|---|---|---|---|
| 1 | 1 to 5 | 20 to 44.7 | Direct connection of the I/O to the environment (e.g. LIN) |
| 2 | 0.1 to 0.5 | 6.3 to 14.1 | Direct connection of the I/O to the environment, but some R-L-C low-pass filtering is available (e.g. sensor interfaces) |
| 3 | 0.01 to 0.05 | 2 to 4.5 | No direct connection of the I/O to the environment (e.g. interfacing with IC's mounted on the same module) |

is depicted in Fig. 2.8. The nature of this measurement makes it very suitable to be incorporated directly in circuit simulators. The DPI specification further states that a forward power is injected through the coupling block in the IC pin no matter whether it is reflected or absorbed. The forward injected power level depends on the application of the IC and on the IC pin itself: a summary is presented in Table 2.2 [Mey03]. Observe that the relation between the forward injected EMI power ($P_{f\_EMI}$) and the EMI source voltage amplitude ($V_{emi}$) is expressed as follows:

$$V_{emi} = 2 \cdot \sqrt{2} \cdot \sqrt{P_{f\_EMI} \cdot R_s} \tag{2.13}$$

For many integrated applications, the DPI measurements are used for EMC pre-compliance testing.

# Chapter 3

# EMC of Integrated Circuits versus Distortion

*The Wall Street Journal reports that military investigators are exploring the possibility the electromagnetic interference may have been the cause of two friendly fire incidents during the Iraq war involving Patriot missiles that resulted in downing of two allied fighters and the deaths of three airmen. According to the Journal report, investigators have ruled out either manual error by the operators of the Patriot missile batteries, or mistakes by the missiles themselves, and are now focusing on whether the extremely close positioning of multiple missile batteries on the ground resulted in elevated levels of EMI that interfered with the system's high-powered radars.*

—Quoted from [Arm07]

## 1 Introduction

As has been highlighted in Chap. 2, numerous EMI sources can be distinguished and classified according to their origin. These EMI sources create conducted disturbances in PCB tracks, and enter the integrated electronic circuits via all possible electromagnetic paths. The majority of these paths are obvious and well identified (e.g. the pins connecting an IC to the outside world, as well as the resulting inductive and capacitive crosstalk which may exist between neighboring pins). However, the exact relationships taking into account the overall coupling between all parasitic elements, the effect of bonding wires, various package types etc., are still not accurately evaluated. The sum of all these phenomena makes the EMC problem at IC level especially difficult to grasp and extremely complex to analyze [Wie06].

External precautions can be taken at system level to ensure that the IC's are well shielded from disturbances, as is described in [Goe01]. Shielding, or screening the whole appliance by means of a Faraday cage is an effective way of keeping unwanted disturbances outside, as well as generated radiated and induced emissions inside in order to prevent an interference aspect elsewhere

J.-M. Redouté, M. Steyaert, *EMC of Analog Integrated Circuits*,
Analog Circuits and Signal Processing,
DOI 10.1007/978-90-481-3230-0_3, © Springer Science+Business Media B.V. 2010

in the system: basic theoretical shielding concepts and formulas are reviewed in [Swa94]. The power supply can be filtered, and electrical separators (e.g. isolation transformers and optocouplers) may be used to block undesirable conducted common mode disturbances while conveying the wanted differential mode signals. Special protective components suppress very short transients (e.g. like those caused by lightning or by ignition sparks). Finally, inductive and capacitive crosstalk can be mitigated by a careful system layout and by ensuring a proper cabling. As a rule, proper grounding while keeping in mind that all current flows in loops solves many (seemingly unsolvable) EMC problems. Finally, a careful PCB design is critical to prevent EMC related problems in an IC which is connected to a PCB board [Arc04, Goe01]. In spite of previously enumerated external precautions, they may not be sufficient to guarantee a faultless IC operation under all possible EMC conditions. Additionally, external components increase the cost associated to the bill of material (BOM), and so the usage of filtering and protective devices like chokes, shields, etc. is limited to the bare minimum required [Mey03]. Moreover, in many design flows, the PCB fabrication is fully separated from the IC design, and so the IC should be able to meet the stringent EMC requirements regardless of the PCB it's ultimately connected to. The latter is especially important for IC manufacturers, which are very often required to provide EMI resisting IC's independently of the PCB, system or end product where these are going to be used.

This chapter describes the relationship between EMI in IC's and distortion. Once an EMI disturbance manages to reach an internal circuit node in an integrated circuit, it mixes with the wanted signal(s) and induces distortion in that node. In the event that the EMI signal is distorted nonlinearly, it will contain harmonic components, intermodulation products and an undesirable DC component. While the harmonic and intermodulation components may distort the wanted signal(s), the generated DC component may be accumulated, generating a shift in the DC bias operating point, in case the circuit's bandwidth lies below the largest EMI induced harmonics and intermodulation products. In the latter case, the circuit's operating point(s) is (are) forcibly pulled out of its (their) correct bias region. Since the origin of DC shift lies in the accumulation of an asymmetric nonlinearly distorted signal, it is important to define and illustrate thoroughly this phenomenon. However, although theoretical developments are mandatory in order to understand appearing EMI induced phenomena, they do not directly provide more insight as to how a simple circuit may be disturbed by EMI, or even forced out of its operation region when a relatively small and apparently harmless EMI signal is injected into one of its circuit nodes. For this reason, following a brief classification of existing distortion types, the general concepts and observations are derived using four basic case studies, namely:

- **Case study 1: a diode connected NMOS transistor**, where the DC shift effect is introduced and its relation with rectification is established analytically.
- **Case study 2: a NMOS source follower**, where it is illustrated that a classic linear analysis is not necessarily sufficient in order to predict EMC phenomena like DC shift accurately. In the same sense, this circuit illustrates that EMI induced DC shift may occur much more often than might be expected at first.
- **Case study 3: a NMOS current mirror**, where the appearing EMI problems are derived mathematically; afterwards, these observations are applied in the design of two possible EMI resisting current mirror topologies.
- **Case study 4: EMI susceptibility in ESD protections**, where the different nonlinearity types and their impact on DC shift are described.

These four circuits help to define and clarify the EMC problem in analog integrated circuits using a bottom-up approach. As will become apparent in the course of this chapter, even very small and very basic analog circuits can (and will) behave erratically the moment they are disturbed by EMI.

## 2 Relationship between EMI resisting design and distortion

Distortion is a common phenomenon in integrated electronics: although the topic itself is well documented, it remains a distrusted subject as well as continuous source of concern during the design of analog integrated electronics. As cited in [Wam98], distortion is nothing else but a deviation of the output signal from the wanted waveform. Distortion occurs in linear circuits (linear distortion) as well as in nonlinear ones (nonlinear distortion). When conducted EMI is injected into an arbitrary integrated circuit through one or more pins, it obviously introduces a certain amount of distortion. Keeping in mind that EMI does not necessarily follow the signal path and that it may couple through and between any parasitic path leading to an outside pin, existing distortion analyses techniques are applicable as such when designing EMI resisting integrated circuits. As illustrated further on, different distortion types each cause a different EMC circuit behavior. To this end, linear and nonlinear distortion is considered separately.

### 2.1 Linear distortion

Linear distortion is the distortion which arises in purely linear circuits, as soon as one or more linear components exhibit a non-flat frequency response [Wam98]. Consider as an example a square wave which is applied at the input of a $R$-$C$ low-pass filter: the output of this $R$-$C$ filter is linearly distorted, because the high frequency sinusoidal components are more attenuated than the

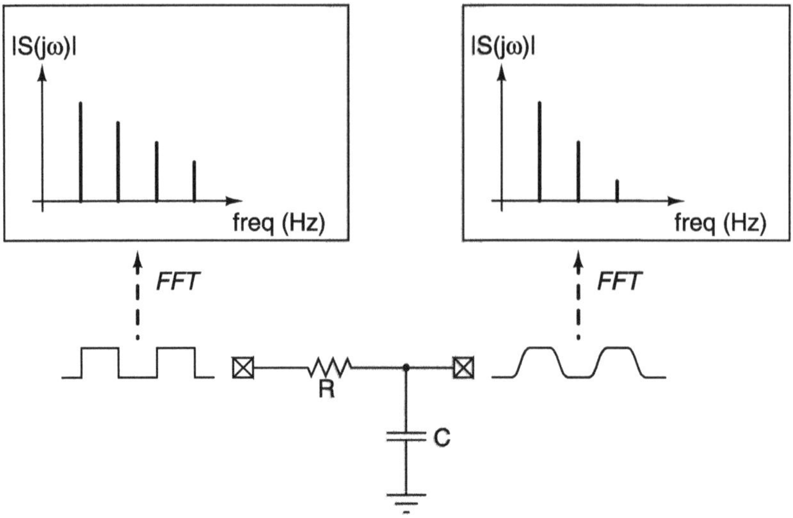


*Figure 3.1.* Linear distortion in a *R-C* low-pass filter.

low frequency ones (Fig. 3.1). However, no new spectral lines are created in the frequency spectrum: this is the basic characteristic of linear distortion [San99]. Linear distortion also appears in any practical amplifier owing to the non-ideal gain and phase variations as a function of the frequency. When EMI is injected in a fully linear circuit, it behaves no differently than any other wanted signal: as such, the interfering signal is linearly distorted in the event that it has frequency components which lie above the circuit's cut-off frequency. This linearly distorted EMI signal is then superposed on the wanted signal(s) which are processed by this circuit, hereby causing an unwanted ripple. This ripple may distort the amplitude of wanted signals and may equally impair the correct circuit's behavior (e.g. by triggering false states in digital circuitry). Much more importantly, this ripple may couple to neighboring circuits which may in turn exhibit a nonlinear behavior, causing nonlinear distortion as described in the next section.

## 2.2 Nonlinear distortion (rectification)

When the main parasitic paths through which the EMI couples in a particular integrated (sub)circuit are identified, measures can be taken in order to filter the resulting EMI induced ripple. Decoupling capacitors, linear filters and other circuit techniques (like using opamps with a high power supply rejection ratio in order to shield the wanted signals from electromagnetic noise which is present on the power supply rails) must be used to filter the EMI before it reaches and mixes with sensitive and nonlinear circuit nodes. Failing to do so results in nonlinear distortion [Fio03]. Nonlinear distortion arises in nonlinear circuits, and amounts to the distortion of the signal amplitude as well as to the position of spectral components. Two different nonlinear distortion types are identified: harmonic and intermodulation distortion. Both are derived and explained here below.

- **Harmonic distortion:** Consider a memoryless, weakly nonlinear system, of which the output signal ($v_o$) is related to the input signal ($v_i$) as follows [Pap99]:

  $$v_o = a_1 v_i + a_2 v_i^2 + a_3 v_i^3 + \cdots \tag{3.1}$$

  Assume that the input signal is a sinusoidal EMI signal, expressed as follows:

  $$v_i = \hat{v} \cdot \sin(\omega t) \tag{3.2}$$

  Substituting (3.2) in (3.1), and performing basic trigonometric operations yields:

  $$v_o = \left(\frac{a_2\hat{v}^2}{2} + \frac{3a_4\hat{v}^4}{8} + \cdots\right) + \left(a_1\hat{v} + \frac{3a_3\hat{v}^3}{4} + \cdots\right)\sin(\omega t) - \left(\frac{a_2\hat{v}^2}{2} + \frac{a_4\hat{v}^4}{2} + \cdots\right)\cos(2\omega t) - \cdots \tag{3.3}$$

  Equation (3.3) illustrates that when nonlinear circuits are excited with a single sinusoidal signal, the frequency spectrum of the output contains a spectral component at the original (fundamental) frequency, as well as spectral components at multiples of the fundamental frequency (harmonic frequencies). This type of distortion in commonly referred to as harmonic distortion, since the distortion components manifest themselves at harmonics (multiples) of the fundamental frequency [San99]. Harmonic distortion is particularly harmful because the harmonic components associated to the nonlinear distortion of a sinusoidal out-of-band EMI signal, may appear in the signal band, even if the EMI frequency band is not interfering with the wanted signal band. From then on, filtering or removing interfering EMI harmonic component(s) becomes very difficult.

  Moreover, observe in (3.3) that a component at DC appears as well. This DC component depends on the even-order nonlinear behavior, as calculated in [Wam98]: this is not very surprising, since even-order harmonics are related to asymmetrical behavior (resulting in a shift of the DC value). This DC component constitutes a serious concern for EMI resisting circuit design. Indeed, the DC shift phenomenon which arises when this DC component is accumulated (e.g. in a capacitor), is extremely harmful because the correct DC operating region of a given circuit may radically change under influence of an interfering EMI signal: in extremis, particular circuit nodes as well as subsequent stages may be forced into saturation or complete cut-off. This process of accumulating the DC component is called *charge pumping* [Red05], while DC shift is the result of the shift in DC bias. Because DC shift is a DC effect, it is not possible to filter or simply nullify it once it has taken place. Consequently, in order to increase the immunity of the IC in question, two approaches can be followed.

- First, the EMI disturbance can be filtered in order to prevent it from affecting adversely the correct IC operation. However, it is important to filter EMI in a linear way, meaning that they should be intercepted before reaching and interfering with nonlinear circuit nodes.
- Secondly, the bandwidth of the circuit can be increased, so that it lies above the most significant EMI induced harmonics and intermodulation products, preventing the process of accumulating the DC value.

Nonlinear distortion is equally identified as *rectification*: this term originated in the first radio detectors that used a nonlinear element (like a small piece of galena crystal) to rectify an AM modulated radio signal [Phi80]. Two types of rectification are commonly distinguished in the literature: soft and hard rectification [Wie06]. Soft rectification means that the DC operating point shift is not large enough to fully cut-off the device, while hard rectification periodically cuts of the device when EMI is injected into the circuit node in question. This corresponds respectively to the weak and strong nonlinear distortion, as defined further on [Wam98, San99].

- **Intermodulation distortion:** When the EMI injection is a complex waveform which can be represented by the sum of multiple sine waves, or in case it sums with the wanted signal(s), all the sinusoidal frequency components mix and interfere with each other, which results in intermodulation products appearing in a large portion of the frequency spectrum. This troublesome effect is illustrated as follows. Assume that the input of the nonlinear system described mathematically in (3.1) consists of the sum of two sine waves lying at different frequencies, represented as follows:

$$v_i = \hat{v}_1 \cdot \sin(\omega_1 t) + \hat{v}_2 \cdot \sin(\omega_2 t) \tag{3.4}$$

The output of the nonlinear system is now equal to:

$$\begin{aligned} v_o = {} & a_1(\hat{v}_1 \cdot \sin(\omega_1 t) + \hat{v}_2 \cdot \sin(\omega_2 t)) \\ & + a_2(\hat{v}_1 \cdot \sin(\omega_1 t) + \hat{v}_2 \cdot \sin(\omega_2 t))^2 \\ & + a_3(\hat{v}_1 \cdot \sin(\omega_1 t) + \hat{v}_2 \cdot \sin(\omega_2 t))^3 + \cdots \end{aligned} \tag{3.5}$$

Consider first the second-order term in (3.5):

$$a_2 v_i^2 = \frac{1}{2} a_2 (\hat{v}_1^2 + \hat{v}_2^2) - \frac{1}{2} a_2 \hat{v}_1^2 \cos(2\omega_1 t) - \frac{1}{2} a_2 \hat{v}_2^2 \cos(2\omega_2 t)$$
$$- a_2 \hat{v}_1 \hat{v}_2 \cos((\omega_1 + \omega_2)t) + a_2 \hat{v}_1 \hat{v}_2 \cos((\omega_1 - \omega_2)t) \quad (3.6)$$

Clearly, if only the second-order term of the nonlinear system transfer function is considered, the output consists of a DC term, the second-order harmonics of both input signals and finally, two components containing the sum and difference of the two input frequencies respectively. These two last terms are identified as the second-order intermodulation products ($IM_2$). The third-order term of (3.5) is equal to:

$$a_3 v_i^3 = a_3 \hat{v}_1^3 \sin^3(\omega_1 t) + 3 a_3 \hat{v}_1 \hat{v}_2^2 \sin(\omega_1 t) \sin^2(\omega_2 t)$$
$$3 a_3 \hat{v}_1^2 \hat{v}_2 \sin^2(\omega_1 t) \sin(\omega_2 t) + a_3 \hat{v}_2^3 \sin^3(\omega_2 t) \quad (3.7)$$

Previous relationship illustrates that if only the third-order term of the nonlinear system transfer function is considered, the output consists of both fundamental frequencies, the third-order harmonics of both input signals and last but not least, the third-order intermodulation products ($IM_3$) that are formed by summation and subtraction of one of the two fundamentals with the second harmonic component of the other ($\omega_1 \pm 2\omega_2$ and $2\omega_1 \pm \omega_2$). Observe that there is no DC component in the third-order term, since it is generated by even-order nonlinear behavior [Wam98]. Intermodulation may mix two or more sinusoidal out-of-band signals, and warp (intermodulate) them into the signal band: this can for instance be experienced when GSM signals emanating from a cell phone are picked up and demodulated by a neighboring audio amplifier. Although both appliances work at very different frequencies, the GSM signals are intermodulated by the nonlinearities in the audio amplifier and occasion intermodulation components in the audio frequency band, generating the quite recognizable repetitive sound in e.g. computer speakers.

## 2.3 Weak and strong nonlinear distortion

When considering nonlinear distortion, it is important to make a distinction between weak and strong nonlinear distortion. In a very general way, weak nonlinear distortion can be represented in terms of Taylor power series (in case of weakly nonlinear, memoryless or anhysteretic circuits) or Volterra series (in case of weakly nonlinear hysteretic circuits) [Pap99]. The latter requires more extensive calculations compared to the former: however, mathematical relationships exist between both methods so that the Volterra transfer functions can be obtained from the generalized power series expansion of a nonlinear system [Ste83]. The similarity between Volterra and Taylor series is expected, since Volterra series are essentially Taylor power series with a memory function. The Volterra series limited convergence forms the basic limitation of this method, and consequently it may be circumvented using orthogonal functions, Wiener functionals and other complex mathematical techniques [Sch80]: these however, increase the mathematical complexity and fall out of the scope of this work.

A nonlinear system, whose response for a given excitation can be represented by a converging Volterra series, is said to behave in a weakly nonlinear way. A more restrictive definition is offered in [Wam98], and states that *a circuit behaves weakly nonlinearly if, for the applied input signal, it can be accurately described by the first three terms of its (converging) Volterra series.* In practice, this means that the weak nonlinearity is described by the linear signal component together with its lowest even- and/or odd-order distortion term. The weak nonlinear behavior is caused by the curvature of the active devices in their operating region. As an example, in [Fio03], an analytical model of a CMOS operational amplifier which is excited by EMI at its input terminals is developed using Volterra kernels. Measurements confirm that the model predictions are close to experimental results as long as the EMI induced distortion of the MOS input differential pair transistors is weak, namely, as long as these transistors stay biased in the saturation region at all times.

Clearly, the efficient application of Volterra series analysis for nonlinear systems is limited by a condition of convergence [Dob03]. Volterra series describe nonlinear systems similarly to the way to Taylor series approximate an analytical function. In the same way, if a nonlinear system is excited by a small amplitude signal, its Volterra series can be broken down after a few terms. At a certain point, for high EMI amplitude signals, more terms are needed to describe the system properly. For very high amplitudes, the Volterra series diverges, just like the Taylor series, and does no longer represent the behavior of the nonlinear system correctly: this is the case for strong nonlinear distortion. Strong nonlinear behavior is the nonlinearity which is generated when active devices are brutally switched on and off (e.g. clipping appearing at the output of an amplifier which is excited by an input signal exceeding the amplifier input dynamic range). It is in general not possible to obtain accurate closed form expressions of circuits behaving in a strong nonlinear way [Wam98]. Because of the mathematical complexity inherent to Volterra series, memoryless power series are used to describe the nonlinear behavior of circuits in this work. This is justified by the fact that the used calculations concentrate solely on qualitatively identifying the EMC problems appearing on respective circuit nodes and components, in order to derive and design circuit topologies with an improved electromagnetic immunity. The calculations therefore do not focus on the magnitudes of the appearing EMI induced nonlinearities, but are merely used to track down and suppress the main causes for the circuit's electromagnetic susceptibility.

In practice, circuits are seldom fully linear: however, they may be considered as behaving linearly as long as the signals injected into them are small (typically, as long as the harmonic components and intermodulation products stay below the noise floor) and as long as they stay biased in their correct operating region. This is the same as saying that when the EMI amplitude is small, power (and Volterra) series may be deftly approximated by their first (linear) term. Consequently, when dealing with the injection of EMI, it is of paramount importance to minimize the injected EMI signal amplitude as much as possible, before it reaches a nonlinear circuit node. The smaller the amplitude of a signal reaching a nonlinear node, the smaller the experienced curvature of the active device and, as a consequence, the better the linearity. Once the EMI is interfering with a nonlinear circuit node, the disturbances start to mix and intermodulate, and as a result, pollute a large portion of the frequency band. Even worse, self-mixing generates a DC component, which may alter the correct biasing and may therefore prohibit the circuit from functioning correctly, if its bandwidth lies below the EMI induced frequencies. These effects have been studied and reported for bipolar transistors as well as for MOS transistors [Ric79b, For79].

As mentioned previously, two approaches are possible to prevent DC shift from occurring. First of all, the EMI may be contained before it reaches a nonlinear circuit node: paradoxically, filtering a nonlinearly distorted EMI signal may even yield worse results, as is illustrated further on. The second possible approach is to increase the bandwidth of the circuit, hereby preventing accumulation. These theoretical observations are illustrated with four basic case study examples in the following sections.

## 3 Case study 1: diode connected NMOS transistor

Consider a diode connected NMOS transistor, which is biased by a DC current source $I_{IN}$ (Fig. 3.2a). Assuming that the NMOS transistor is biased in strong inversion, and using first order MOS transistor formulas, the gate-source voltage of this transistor is equal to [San06]:

$$V_{GS} = \sqrt{\frac{I_{IN}}{\frac{\mu C_{ox}}{2} \cdot \frac{W}{L}}} + V_t \tag{3.8}$$

Assume that an EMI AC current $i_{emi}$ is superposed on the DC voltage of $I_{IN}$. The interference $i_{emi}$ is sinusoidal, and defined as follows:

$$i_{emi} = \hat{i} \cdot \sin(\omega t) \tag{3.9}$$

The total input current is then represented as:

$$I_{in} = I_{IN} + i_{emi} \tag{3.10}$$

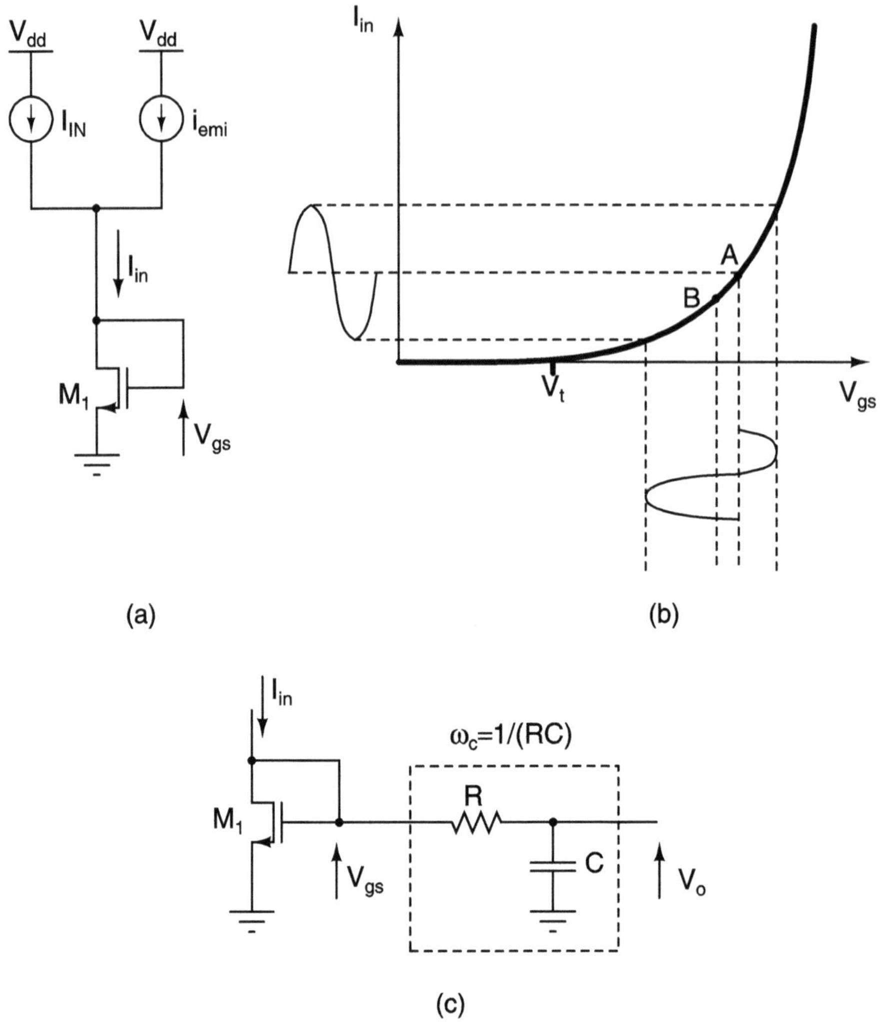


*Figure 3.2.* (a) Diode connected NMOS transistor. (b) DC shift in the diode connected NMOS transistor. (c) Diode connected NMOS transistor followed by an ideal low-pass filter, with cut-off frequency $\omega_c$.

The total gate-source voltage is expressed accordingly as:

$$V_{gs} = \sqrt{\frac{I_{IN} + i_{emi}}{\frac{\mu C_{ox}}{2} \cdot \frac{W}{L}}} + V_t \tag{3.11}$$

The modulation index ($m$) is defined as the ratio between the EMI amplitude and the DC bias current:

$$m = \frac{\hat{i}}{I_{IN}} \tag{3.12}$$

As long as $m < 1$, the amplitude of the EMI is smaller than the bias current $I_{IN}$. In that case, the diode connected transistor is always conducting a forward current. The relationship between the amplitude of the EMI signal and the magnitude of the input bias current is then expressed as a function of $m$, meaning that (3.10) is rewritten as:

$$I_{in} = I_{IN} \cdot (1 + m \cdot \sin(\omega t)) \tag{3.13}$$

Substituting (3.13) in (3.11), the following expression for the gate-source voltage is obtained:

$$V_{gs} = V_t + \sqrt{\frac{I_{IN}}{\frac{\mu C_{ox}}{2}\frac{W}{L}}} \cdot \sqrt{1 + m \cdot \sin(\omega t)} \tag{3.14}$$

As long as the modulation index $m$ is smaller than 1, Taylor series can be used to expand expression (3.14) [Gly96]. This yields:

$$\begin{aligned} V_{gs} = V_t + \sqrt{\frac{I_{IN}}{\frac{\mu C_{ox}}{2}\frac{W}{L}}} \cdot \Bigg(1 + \frac{m}{2} \cdot \sin(\omega t) - \frac{m^2}{8} \cdot \sin^2(\omega t) \\ + \frac{m^3}{16} \cdot \sin^3(\omega t) - \cdots \Bigg) \end{aligned} \tag{3.15}$$

Observe that the nonlinear $V_{gs}$ signal has been expanded into a power series, and that it is of the same form as the general expression given in (3.1). The mean value over time of the gate-source voltage is now equal to [Gly96]:

$$\begin{aligned} \overline{V_{gs}} &= \lim_{T \to \infty} \frac{1}{T} \int_{-\frac{T}{2}}^{\frac{T}{2}} V_{gs} \cdot dt \\ &= V_t + \sqrt{\frac{I_{IN}}{\frac{\mu C_{ox}}{2}\frac{W}{L}}} \cdot \left(1 - \frac{1}{16} \cdot m^2 - \frac{15}{1024} \cdot m^4 - \frac{105}{16384} \cdot m^6 - \cdots \right) \end{aligned} \tag{3.16}$$

Previous expression shows that the average value $\overline{V_{gs}}$ shifts downward owing to the EMI. The visual representation of this effect is sketched in Fig. 3.2b. Observe that owing to the EMI disturbance $i_{emi}$, the operating point moves from A to B. At this point, two cases must be distinguished from each other. In order to do this, assume that voltage $V_{gs}$ is filtered by means of a $R$-$C$ low-pass filter, with a cut-off frequency $\omega_c$, as depicted in Fig. 3.2c. Out of simplicity, resistor $R$ is considered as being much higher than $1/g_{m1}$, where $g_{m1}$ represents the transconductance of $M_1$, so as not to load the input node.

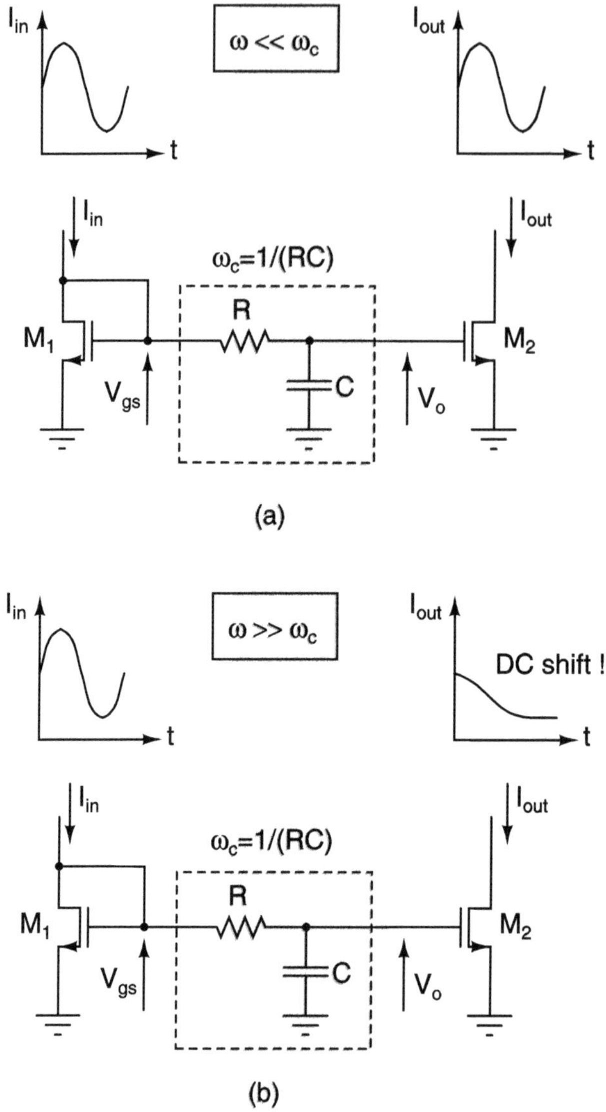


*Figure 3.3.* (a) EMI frequency lies below the low-pass filter cut-off frequency. (b) EMI frequency lies above the low-pass filter cut-off frequency.

- **EMI frequency lies below the filter cut-off frequency** ($\omega \ll \omega_c$)**:** Because of the high bandwidth of the low-pass filter, the output voltage $V_o$ is equal to $V_{gs}$, and expressed as in (3.15). Observe that although $V_o$ is a nonlinear function of the EMI current, no DC shift is taking place, since the reverse operation yields the original EMI current which is injected in the circuit (depicted in Fig. 3.3a). Because the bandwidth of the low-pass filter is much higher than the relevant EMI induced frequencies, no extra charge is accumulated in capacitor $C$.

- **EMI frequency lies above the filter cut-off frequency** ($\omega \gg \omega_c$)**:** Since the EMI induced frequencies are situated above the low-pass filter bandwidth, $V_o$ does not contain high frequency EMI components, and is consequently expressed as in (3.16). In this case, there is a shift of the DC operating point, because the only remaining EMI induced component is the DC component which is accumulated in $C$. It is useful to observe that the reverse operation does not yield the original EMI current which is injected in the circuit (depicted in Fig. 3.3b). EMI induced DC shift is a very serious phenomenon, in the sense that it can fully debias the circuit as well as subsequent circuits. In addition, it indicates an even-ordered nonlinear circuit behavior, as discussed in Sect. 2. Following the previous explanation, there are two possibilities to prevent DC shift. The first is to filter the EMI signal before it reaches the nonlinear circuit node (in this case, by decreasing the EMI contribution in the input current). The second is to increase the bandwidth of the circuit, so that it is larger than the relevant EMI induced harmonic frequencies and intermodulation products (in this case, by increasing the cut-off frequency $\omega_c$).

As specified earlier, the mathematical development which was performed in the first part of this case study is an approximation using first order MOS transistor models. In order to increase the accuracy of the calculations, complex circuit models (such as provided by the BSIM-family) in combination with powerful CAD tools (e.g. SPICE, SPECTRE or ELDO) are used to refine the results obtained with hand calculations. Nevertheless, the latter provide a clear insight in the device operation, which explains their importance: however, they do not take every effect into account. As an example, in this case, they are only valid as long as $m$ is smaller than 1. If $m$ increases above 1, the amplitude of $i_{emi}$ becomes larger than the bias current resulting at a certain point in strongly nonlinear distortion of the drain current. Such excessive overshoots may substantially shorten the lifetime of the IC by exceeding the breakdown voltage, while the heavy undershoots may trigger latch-up: without extra precautions, the latter may introduce a significant substrate current flow via the parasitic bulk-drain diode in the diode connected transistor [Has00]. In this event, Taylor and Volterra series may not be used any longer, and other means to expand this function must be employed: however, this involves a lot of intricate calculations that do not contribute directly to more basic insight nor provide a high degree of accuracy. Suffice it to say that for higher values of $m$, a strong rectification occurs.

## 4 Case study 2: NMOS source follower

Consider a common-drain stage, depicted in Fig. 3.4. Source resistor $R_S$ biases the source follower, and is shunted by a decoupling capacitor $C_S$. An AC EMI source $v_{emi}$ is conveyed to the gate of the transistor. This section illustrates that DC shift may occur in this circuit, depending on the values of $R_S$ and $C_S$. It will also be shown that although small signal analyses are quite straightforward to use in order to detect potential EMC issues, they are in themselves not sufficient to quantize EMC problems because they do not take nonlinear phenomena (like nonlinear distortion) into account. Shortly, this signifies that large signal calculations and simulations are still very necessary to check the exact susceptibility of a particular circuit to EMI. This is now illustrated with the source follower example. Assume that the EMI disturbance is small and sinusoidal, and expressed as follows:

$$v_{emi} = \hat{v} \cdot \sin(\omega t) \tag{3.17}$$

The total drain-source current flowing through $M_1$ can then be calculated:

$$I_{ds} = \frac{\mu C_{ox}}{2} \cdot \frac{W}{L} \cdot (V_{GS} + v_{gs} - V_t)^2 \tag{3.18}$$

In previous expression, $V_{GS}$ represents the DC gate-source bias voltage of transistor $M_1$. In case EMI is present and if the EMI amplitude is small enough so that linear analysis can be used, the eigenfunction property of linear time invariant systems can be applied [Dut97]. The AC gate-source voltage ($v_{gs}$) is then expressed using the inverse Laplace operator and substituting (3.17) [Dut97]:

$$v_{gs}(t) = \mathcal{L}^{-1}\{H(s) \cdot v_{emi}(s)\} = |H(j\omega)| \cdot \hat{v} \cdot \sin(\omega t + \phi) \tag{3.19}$$

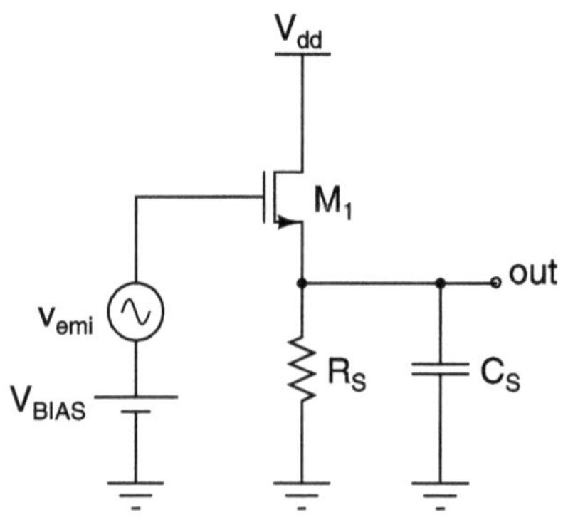


*Figure 3.4.* Source follower with EMI injection at the gate.

where $|H(j\omega)|$ and $\phi$ represent respectively the magnitude and the phase of transfer function $H(s)$, expressed as:

$$H(s) = \frac{v_{gs}(s)}{v_{emi}(s)} = \frac{\frac{1}{R_S} + s \cdot C_S}{g_m + \frac{1}{R_S} + s \cdot C_S} \tag{3.20}$$

In practice, source resistance $R_S$ is much larger than $1/g_m$, and so for higher frequencies, transfer function $H(s)$ is simplified to:

$$H(s) = \frac{v_{gs}(s)}{v_{emi}(s)} \approx \frac{s \cdot C_S}{g_m + s \cdot C_S} \tag{3.21}$$

The average drain-source current flowing through $M_1$ is obtained by time-averaging expression (3.18) [Gly96]. This yields:

$$\begin{aligned} \overline{I_{out}} &= \lim_{T \to \infty} \frac{1}{T} \int_{-\frac{T}{2}}^{\frac{T}{2}} I_{out}(t) \cdot dt \\ &= \overline{\frac{\mu C_{ox}}{2} \cdot \frac{W}{L} \cdot (V_{GS} - V_t)^2} + \overline{\mu C_{ox} \cdot \frac{W}{L} \cdot (V_{GS} - V_t) \cdot v_{gs}(t)} \\ &+ \overline{\frac{\mu C_{ox}}{2} \cdot \frac{W}{L} \cdot v_{gs}^2(t)} \end{aligned} \tag{3.22}$$

Previous expression consists of three distinct terms. The first term is the wanted DC term: it is the expression for the DC bias current itself without the presence of EMI. The second term represents a current component that is linearly proportional to the sinusoidal EMI signal, and is commonly referred to as the linearized transconductance ($g_m$) term: its average is therefore zero. The third term is proportional to the square of the EMI signal: its average value is not equal to zero, and so this term is responsible for DC shift. Substituting expression (3.19) in (3.22), the DC shift of the output current is equal to [1]:

$$\overline{I_{out\_DCshift}} = \frac{\mu C_{ox}}{4} \cdot \frac{W}{L} \cdot (|H(j\omega)| \cdot \hat{v})^2 \tag{3.23}$$

Observe that the DC shift can be reduced by decreasing the EMI signal, and by decreasing $|H(j\omega)|$. As is apparent from (3.21), this can be realized by

[1] Strictly speaking, since DC shift is generated by the accumulation of an asymmetrically rectified signal, it is not correct at this point to speak of a "DC shift of the output current", since the accumulation has, properly speaking, not taken place yet. An appropriate nametag would sound like "the rectified output current contribution, whose accumulated value yields a DC shift". This would however needlessly complicate the explanations. For this same reason, the quadratic term resulting from expansions which are similar to (3.22), will equally be referred to as the "DC shift term" in the remainder of this work. In the same sense, when talking about DC shift, accumulation will always be assumed implicitly.

increasing $g_m$ and decreasing $C_S$. In fact, for a very high transconductance, this DC shift term becomes so small that it is close to negligible.

This approach is valid as long as the amplitude of the EMI signal is small, and as long as $M_1$ stays in saturation. If the EMI amplitude is increased, $M_1$ is not in saturation any longer during the full length of the EMI period and this induces strong nonlinear distortion. Strange as it may seem, no large EMI amplitudes are even required to reach this situation because of the DC shift that occurs on the source of $M_1$ for high EMI frequency signals. This DC shift does not appear directly in the small signal analysis because of the intrinsic linearization that the latter implies, but has nevertheless a crucial impact on the circuit's EMI behavior, as is illustrated here. If the EMI amplitude is increased so that a small signal analysis is no longer applicable, a large signal approach must be used. Two situations may be distinguished in this event [Wie06]:

- **The cut-off frequency formed by $C_S$ and $R_S$ is larger than the EMI frequency:** the source of $M_1$ charges the output node, while source resistor $R_s$ discharges this node. Since the time constant formed by $C_S$ and $R_S$ is sufficiently small, $C_S$ is (almost) fully discharged within every sine wave period. Therefore, no rectification is taking place.

- **The cut-off frequency formed by $C_S$ and $R_S$ is smaller than the EMI frequency:** the source of $M_1$ still charges the output node, while source resistor $R_s$ discharges this node. However, the time constant formed by $C_S$ and $R_S$ is now significant, meaning that all the charge that is stored in $C_S$ is not fully discharged in $R_S$ within one EMI signal period. The charge across the output node is therefore pumped to a higher value. This is comparable to the slew rate phenomenon, which defines the maximum rate of change of the output voltage of an operational amplifier. Likewise, slew rate is generated by the dominant capacitor which is charged and discharged by finite currents [Raz01].

This phenomenon is observed in Fig. 3.5. Observe that DC shift moves into the direction of cut-off of the nonlinear device that has originated it in the first place [Wie06]. The conclusion is that a small signal analysis alone is not sufficient into predicting DC shift: even when considering relatively small EMI signals, strong nonlinear distortions may be experienced, depending on the exact nonlinear behavior of a circuit which is subjected to EMI.

## 5 Case study 3: NMOS current mirror

Consider a classic integrated current mirror, consisting of two NMOS transistors, whose purpose is to provide an arbitrary DC bias current to an integrated circuit [Gra01]. An external DC current source (e.g. a resistor connected to the fixed supply voltage or a current source) determines the amount

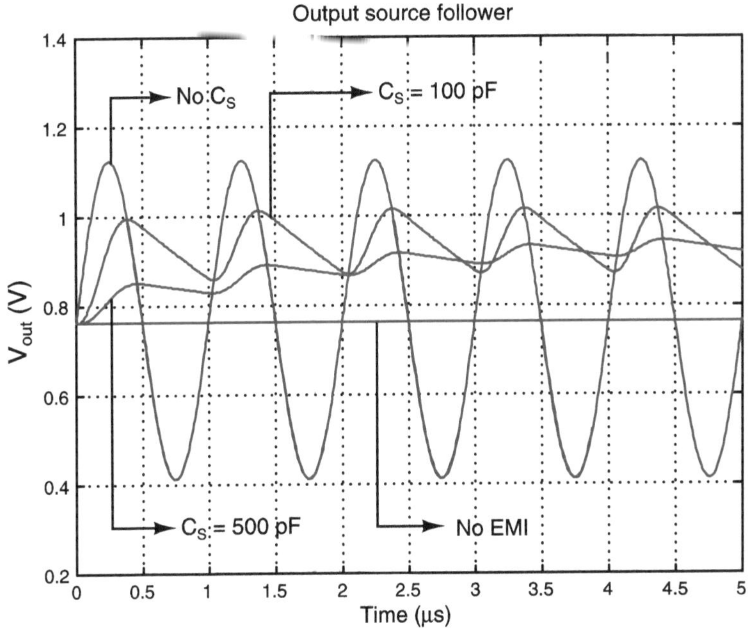


*Figure 3.5.* Output voltage of the source follower.

of input DC bias current (Fig. 3.6a). The current gain transfer function between the input and the output current is set by the $W/L$ ratio of both transistors [San06]. If an out-of-band EMI signal is present on the external net (e.g. on the PCB track connecting $I_{IN}$ and the IC pin), the total current through the first branch of the mirror is represented as the sum of the wanted DC current $I_{IN}$, and the unwanted EMI AC current, called $i_{emi}$ as in (3.10). The interference $i_{emi}$ is further coupled to the output current $I_{out}$, and consequently to the subsequent stages which are biased by this current mirror. This is clearly seen in Figs. 3.7 and 3.8 for respectively small and large EMI signals. Both plots depict $I_{out}$ for the following design example: the DC bias current $I_{IN}$ is 10 μA, and both transistors are equal in size ($W_1/L_1 = W_2/L_2 = 10$ μm/1 μm; $g_{m1} = g_{m2} = 140$ μS). This circuit was simulated in a standard CMOS 0.35 μm technology. Not surprisingly, $I_{out}$ is more disturbed by EMI as the amplitude of $i_{emi}$ increases (refer to Fig. 3.7). The moment the EMI amplitude exceeds the bias current, $V_{gs1}$ is clipped and $I_{out}$ is hereby heavily distorted because $M_1$ is drawn into cut-off (refer to Fig. 3.8).

Externally, various precautions as well as protective and decoupling devices can be foreseen to filter and block $i_{emi}$. However, aside from the extra cost associated to an increased bill of material, an application does not always allow the use of such devices, nor does it necessarily tolerate the presence of a large

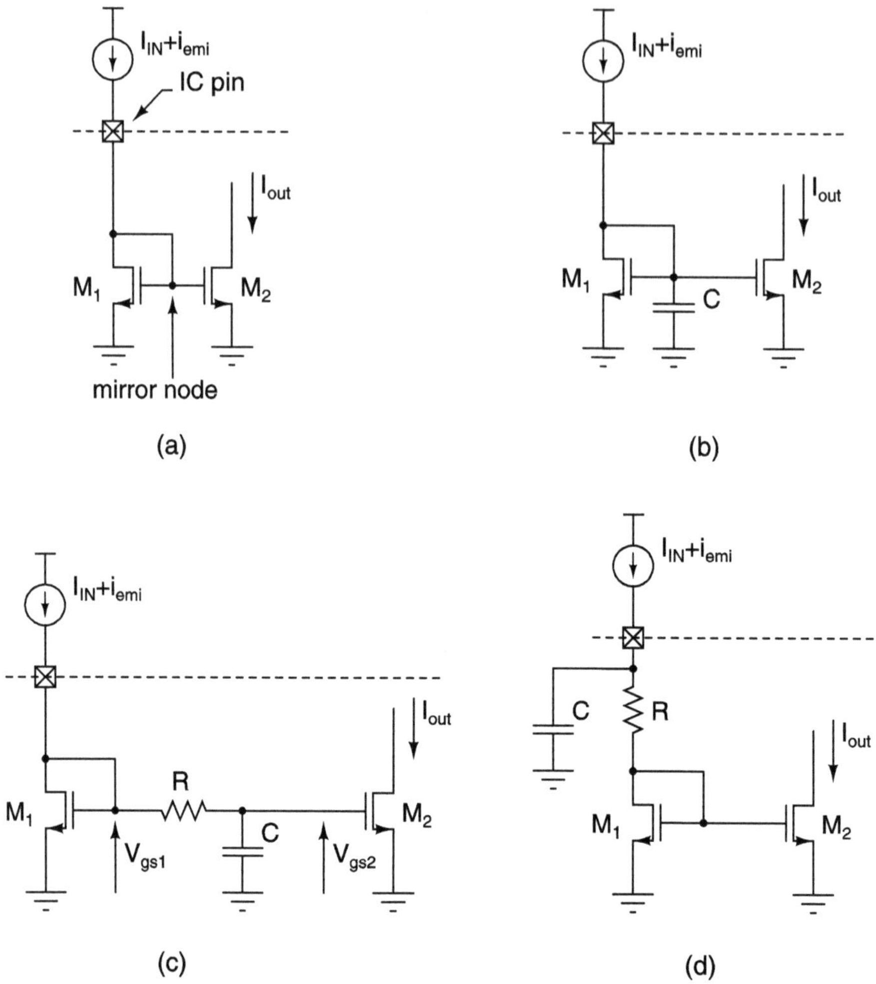


*Figure 3.6.* Classic current mirror (a) EMI injection. (b) With a capacitor between gate and ground. (c) With a low-pass $R$-$C$ filter between the gates. (d) With a low-pass $R$-$C$ filter in series with the input transistor.

decoupling capacitor at an IC pin (e.g. if an in-band wanted signal is present). In addition, this external component may not offer a sufficient EMI filtering throughout the full EMI frequency range. For instance, a large and external decoupling capacitor may be effective at low EMI frequencies, but will not effectively filter very high EMI frequencies owing to its parasitic equivalent series resistor (ESR) and inductor (ESL). For the above reasons, the following analysis considers that since the current mirror circuit is inherently susceptible to EMI which is injected in its input, an external decoupling capacitor is either absent, either simply ineffective at the respective EMI frequencies.

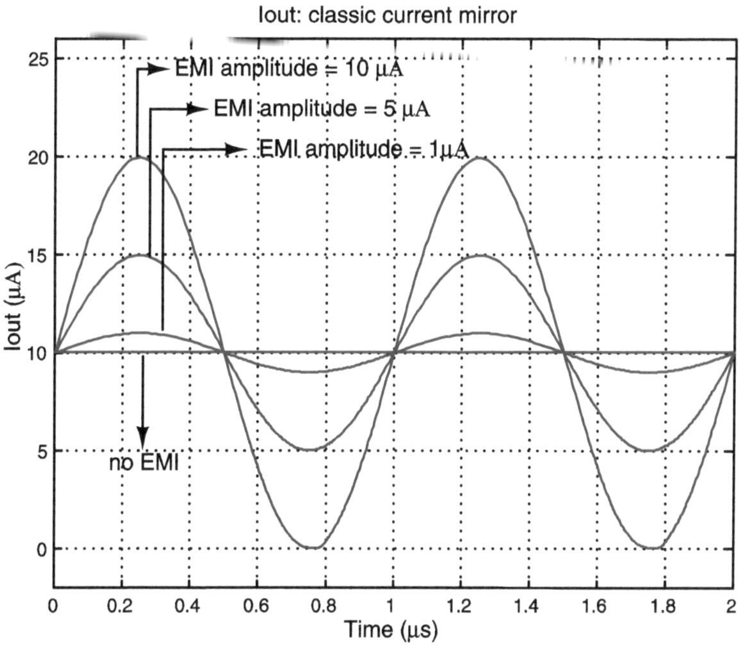


*Figure 3.7.* Output current $I_{out}$, which is heavily contaminated by the EMI for small EMI signals ($m < 1$). However, observe that there is (quasi) no rectification, since the Early effect is small.

Consequently, some internal protection and EMI filtering must be provided internally in the current mirror itself in order to eliminate the disturbing EMI frequencies before they manage to propagate to the output node.

As explained in Sect. 2, small EMI signals trigger weak nonlinear circuit behavior because of the curvature which is characteristic for active devices: this is the case as long as the amplitude of $i_{emi}$ is smaller than the bias current. Referring to the definition of the modulation index $m$ introduced in (3.12), this corresponds to the situation when $m < 1$. It will now be examined if DC shift can occur in this case. The drain currents of $M_1$ and $M_2$ can be expressed as follows using a first order approximation [San06]:

$$\begin{cases} I_{in} = I_{d1} = \frac{\mu C_{ox}}{2} \cdot \frac{W_1}{L_1} \cdot (V_{gs1} - V_t)^2 \cdot (1 + \lambda \cdot V_{ds1}) \\ I_{out} = I_{d2} = \frac{\mu C_{ox}}{2} \cdot \frac{W_2}{L_2} \cdot (V_{gs1} - V_t)^2 \cdot (1 + \lambda \cdot V_{ds2}) \end{cases} \tag{3.24}$$

With $\lambda = 1/(L \cdot V_E)$, where $V_E$ represents the Early voltage [San06]. Combining (3.24) in one relationship expressing the output current as a function of the input current, and taking the mean value over time, yields:

$$\overline{I_{out}(t)} = \overline{I_{in}(t)} \cdot \frac{W_2/L_2}{W_1/L_1} \cdot \frac{1 + \overline{\lambda \cdot V_{ds2}}}{1 + \overline{\lambda \cdot V_{ds1}}} \tag{3.25}$$

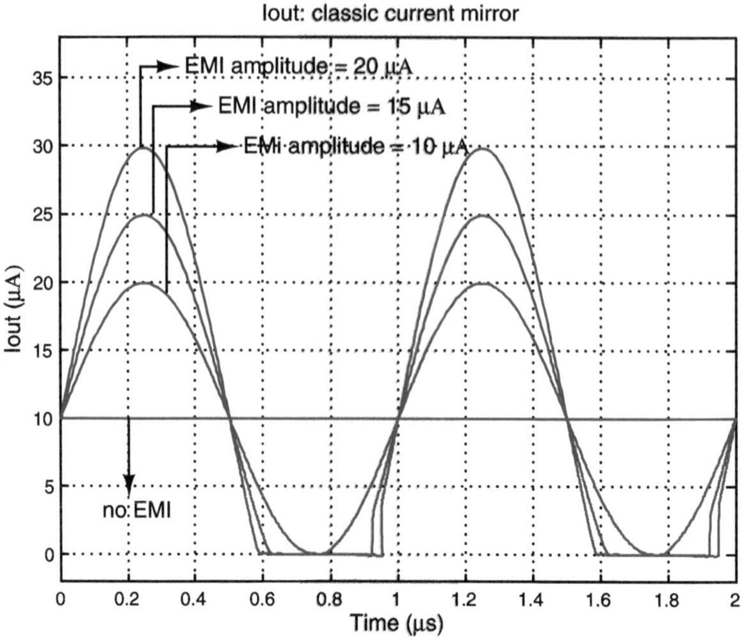


*Figure 3.8.* Output current $I_{out}$, which is heavily contaminated by the EMI for large EMI signals ($m \geq 1$). Since the amplitude EMI disturbance exceeds the bias current, $I_{out}$ is heavily distorted and the average value of $I_{out}$ is shifted upwards.

Under normal circumstances, $\overline{V_{ds2}}$ is roughly equal to its DC value $V_{DS2}$. On the other hand, $V_{ds1} = V_{gs1}$. As long as $m < 1$, (3.16) can be used to approximate $\overline{V_{gs1}}$. Substituting this expression in (3.25) yields the following equality:

$$\overline{I_{out}(t)} = \overline{I_{in}(t)} \cdot \frac{W_2/L_2}{W_1/L_1} \cdot \frac{1 + \lambda \cdot V_{DS2}}{1 + \lambda \cdot V_t + \lambda \cdot \sqrt{\frac{I_{IN}}{\frac{\mu C_{ox}}{2}\frac{W}{L}}} \cdot (1 - \frac{1}{16} \cdot m^2 - \cdots)} \tag{3.26}$$

Previous equation shows that a small amount of DC shift can possibly occur owing to the Early effect. However, it is quite small since it is multiplied by $\lambda$, and it will therefore be neglected in the rest of this section. This is clearly seen in Fig. 3.7, where no apparent rectification is distinguished. $I_{out}$ is consequently expressed as follows:

$$I_{out} = \frac{W_2/L_2}{W_1/L_1} \cdot I_{in} = \frac{W_2/L_2}{W_1/L_1} \cdot (I_{IN} + m \cdot I_{IN} \cdot \sin(\omega t)) \tag{3.27}$$

If the amplitude of $i_{emi}$ increases beyond the bias current ($m \geq 1$), $M_2$ will clip the undershoots, since it can not source current. Evidently, this results in a heavy nonlinear $I_{out}$. In turn, this nonlinearity can induce DC shift in case it is accumulated, as can be appreciated in Fig. 3.8. Clearly, the output current is no longer expressed as in (3.27).

It will be shown later how the developed EMI resisting current mirrors are made much less susceptible to this heavy nonlinear behavior by moving the boundary between weak and strong nonlinear behavior to a higher value. This boundary is referred to "input dynamic range" in the remainder of this section. However, for the time being, the main focus lies on exploring suitable ways to filter the EMI disturbance before it reaches and harms the output node, and this without generating DC shift. Both requirements can be accomplished by reducing the amplitude of $i_{emi}$ before the latter reaches the input node. Next paragraphs illustrate that this constraint is not so easily accomplished as it may seem at first glance. Again, as frequently happens in EMC robust IC design, there's more to the picture than meets the eye.

## 5.1 Capacitor decoupling the mirror node

An internal capacitor between the gate of $M_1$ and ground can be added to filter the EMI (Fig. 3.6b). A small signal analysis yields a current transfer function which is characterized by a real pole at $g_{m1}/C$, with $C$ representing the total capacitance between the gate node and ground, and a right half plane zero due to the parasitic feed-forward capacitance, which can usually be disregarded [Ala97]. Consequently, this capacitor $C$ attenuates EMI frequencies lying beyond the bandwidth frequency of the current mirror, set at $g_{m1}/C$. Unfortunately, since the bandwidth is limited by $g_{m1}$, capacitor $C$ needs to be quite high in order to place this pole below the lowest EMI frequencies. As an example, to obtain an arbitrary attenuation of $-40$ dB at 1 MHz, the mirror pole must be placed at 10 kHz. With $g_{m1}$ equaling 140 μS corresponding to the previous design example, this means that the needed $C$ amounts to 2.2 nF, which is quite a high value to integrate. This fact makes this solution rather unpractical, and therefore not very useful in an integrated environment.

Observe that although $C$ is connected to the mirror node, this capacitor does not cause DC shift if ideal transistors are used and as long as the Early effect is negligible, as in the previous section. This is proven as follows. Kirchhoff's current law applied on the input node of this circuit yields the following equation:

$$I_{in}(t) = I_C(t) + I_{d1}(t) \tag{3.28}$$

Where $I_C(t)$ is the total current flowing through the capacitor, and $I_{d1}(t)$ is the drain-source current flowing through $M_1$. Since both $M_1$ and $M_2$ have the same gate-source voltage, $I_{out}$ is expressed as:

$$I_{out}(t) = \frac{W_2/L_2}{W_1/L_1} \cdot I_{d1}(t) \tag{3.29}$$

Combining (3.29) and (3.28), the average value of $I_{in}$ over time is calculated as follows [Gly96]:

$$\begin{aligned}\overline{I_{in}(t)} &= \lim_{T\to\infty} \frac{1}{T}\int_{-\frac{T}{2}}^{\frac{T}{2}} I_C(t)\cdot dt + \overline{I_{d1}(t)} \\ &= \lim_{T\to\infty} \frac{q_C(\frac{T}{2}) - q_C(-\frac{T}{2})}{T} + \frac{W_1/L_1}{W_2/L_2}\cdot\overline{I_{out}(t)} \\ &= \frac{W_1/L_1}{W_2/L_2}\cdot\overline{I_{out}(t)} \qquad (3.30)\end{aligned}$$

where $q_C(t)$ is the mean net charge stored in the capacitor at time $t$. Since a capacitor does not conduct a net current, the average charge over time which is stored in a capacitor is not altered by EMI. The mean input current is therefore equal to the mean output current, and so there is no charge pumping, except for a negligible amount determined by the Early effect, as derived in (3.26). Disregarding the Early effect, $I_{out}$ is therefore again expressed as a function of the modulation index $m$:

$$I_{out} = \frac{W_2/L_2}{W_1/L_1}\cdot I_{d1} = \frac{W_2/L_2}{W_1/L_1}\cdot I_{IN} + m\cdot H_1(j\omega)\cdot I_{IN}\cdot\sin(\omega t) \qquad (3.31)$$

where $H_1(j\omega)$ represents the current transfer function from the input to the output:

$$H_1(s) = \frac{i_{out}(s)}{i_{emi}(s)} = \frac{g_{m2}/g_{m1}}{1+\frac{s\cdot C}{g_{m1}}} \qquad (3.32)$$

Observe that the EMI component which is flowing through $M_1$ and which is consequently coupled to the output is attenuated by $H_1(s)$. As a results, the circuit operates in its weakly nonlinear region as long as $m\cdot|H_1(j\omega)| < 1$, which is a considerable improvement compared to (3.27), where $m$ had to be below unity in order to stay in the weakly nonlinear region. In other words, the input dynamic range is hereby increased by $1/|H_1(j\omega)|$.

## 5.2 Low-pass $R$-$C$ filter in the mirror node

As mentioned previously, the current mirror depicted in Fig. 3.6b requires a huge capacitance $C$ in order to reduce the current mirror bandwidth below the lowest EMI frequencies. A seemingly possible solution to this problem is to place a low-pass $R$-$C$ filter between the transistor gates, with a cut-off frequency $\omega_c$ that lies significantly lower than the frequency of the EMI disturbance and a large value of $R$ that doesn't load the input node ($R \gg 1/g_{m1}$) (Fig. 3.6c). Evaluating this circuit from a small signal point of view, this solution is satisfactory: the realized current transfer function $H_2(j\omega)$ is now equal to:

$$H_2(s) = \frac{i_{out}(s)}{i_{emi}(s)} = \frac{g_{m2}/g_{m1}}{1 + s \cdot R \cdot C} \tag{3.33}$$

The output current is expressed as follows:

$$I_{out} = \frac{W_2/L_2}{W_1/L_1} \cdot I_{IN} + m \cdot H_2(j\omega) \cdot I_{IN} \cdot \sin(\omega t) \tag{3.34}$$

Since $R$ can easily be made much larger than $1/g_{m1}$, a better EMI filtering results which requires a much smaller capacitor. Unfortunately, the EMI component flowing through $M_1$ is not attenuated, and so the input dynamic range stays small ($m < 1$). In addition, this circuit solution is completely unusable, since the voltage on the mirror node is not a linear function of the input current, and consequently the linear $R$-$C$ filtering generates accumulation. This nonlinear distortion generates harmonics and intermodulation components, and even worse, triggers the detrimental DC shift phenomenon. The latter forcibly alters the correct bias point of the current mirror, forcing it to operate in a different operating region and hereby lowering the average output current value. This effect is derived mathematically as follows. If the interference $i_{emi}$ is modeled as a sinusoidal wave, as in (3.9), and provided that the $R$-$C$ low-pass filter does not load the input node ($R \gg 1/g_{m1}$) and that $m < 1$, the average value of the gate-source voltage of $M_1$ is expressed as in (3.16). As long as the EMI frequency lies above the $R$-$C$ cut-off frequency ($\omega \gg \omega_c$), $V_{gs2}$ can be approximated by the DC value of $V_{gs1}$, and this yields:

$$V_{gs2} \approx \overline{V_{gs1}} = V_t + \sqrt{\frac{I_{IN}}{\frac{\mu C_{ox}}{2}\frac{W_1}{L_1}}} \cdot \left(1 - \frac{1}{16} \cdot m^2 - \frac{15}{1024} \cdot m^4 - \cdots\right) \tag{3.35}$$

The average output current is then equal to:

$$\overline{I_{out}} = I_{IN} \cdot \frac{W_2/L_2}{W_1/L_1} \cdot \left(1 - \frac{1}{16} \cdot m^2 - \frac{15}{1024} \cdot m^4 - \cdots\right) \tag{3.36}$$

This last equation shows that extra terms as a function of $m$ are causing DC shift, since the average output current is no longer equal to the original output current without EMI. Figure 3.9 shows the dramatic effect of DC shift on the output current of the circuit depicted in Fig 3.6c over time, for an EMI signal with a frequency of 1 MHz and different amplitudes (varying from 0 to 20 μA). The bias DC current is 10 μA, and both transistors are equal in size ($W_1/L_1 = W_2/L_2 = 10$ μm/1 μm; $g_{m1} = g_{m2} = 140$ μS). The cut-off frequency of the low-pass filter has been placed at 10 kHz ($R = 100$ kΩ and $C = 160$ pF), in order to provide an arbitrary attenuation of −40 dB at 1 MHz. This circuit was designed and simulated in a standard CMOS 0.35 μm technology.

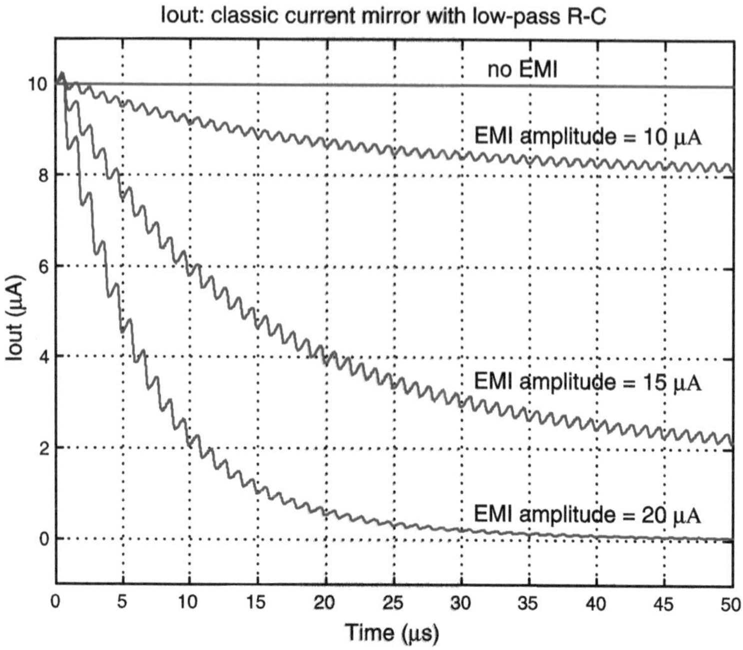


*Figure 3.9.* Dramatic effect of charge pumping on the output current $I_{out}$ of the ordinary current mirror with a low-pass $R$-$C$ filter between the gates. When the disturbing amplitude exceeds the bias current (15 μA and 20 μA), strong nonlinear effects are taking place: these effects (apart from being harmful and susceptible to latch-up) cause even more DC shift.

## 5.3 Low-pass $R$-$C$ filter in the drain of $M_1$

The same $R$-$C$ low-pass filter can be used to filter the EMI before it reaches the nonlinear node: consider Fig. 3.6d. The major disadvantage of this topology is that resistor $R$ increases the required supply voltage, which reduces the practical use of this circuit in present day low supply voltage technologies. As an example, using the same $R$ and $C$ as previously ($R = 100$ kΩ and $C = 160$ pF) and using the same bias current of 10 μA, yields a DC voltage drop of 1 V across the resistor. Since the filtering takes place before the EMI disturbance reaches the nonlinear node, the DC shift is strongly reduced. Furthermore, the input dynamic range is heavily increased, as illustrated here below. The realized current gain transfer function $H_3(j\omega)$ is equal to:

$$H_3(s) = \frac{i_{out}(s)}{i_{emi}(s)} = \frac{g_{m2}/g_{m1}}{s \cdot (C \cdot R + \frac{C}{g_{m1}}) + 1} \tag{3.37}$$

The corresponding output current is now equal to:

$$I_{out} = \frac{W_2/L_2}{W_1/L_1} \cdot I_{IN} + m \cdot H_3(j\omega) \cdot I_{IN} \cdot \sin(\omega t) \tag{3.38}$$

Which means that the input dynamic range is increased by $1/|H_3(j\omega)|$. Because $R$ can be made much larger than $1/g_{m1}$ at the expense of an increased voltage drop, this yields a considerable improvement compared to (3.31).

## 5.4 EMI resisting (4-transistor) current mirror

A current mirror structure countering charge pumping while filtering EMI, is depicted in Fig. 3.10 [Red05]. This circuit bears some similarities with the classic current mirror with "beta helper" using BJT's, which is used to compensate for the base current and the resulting systematic gain error [Gra01]. However, it performs a totally different function, as illustrated here. Transistor $M_2$ isolates the sensitive mirror node from the drain of $M_1$, while $M_3$ completes the DC biasing. The purpose of transistors $M_2$ and $M_3$ is to keep $V_{gs1}$ at a fixed DC level using negative feedback. Capacitors $C_1$ and $C_2$ are not mandatory, but provide the means to integrate a second-order low-pass filter to reduce the EMI contribution in the output current. Observe that the impedance at the mirror node is kept at a very low value by $M_2$ and $M_3$, which ensures the stability of this local feedback loop as long as the impedance at the drain of $M_1$ is kept high. Finally, the local negative feedback loop decreases the (already low) impedance at the mirror node to a very low value, and this property is exploited in full during the design of an EMI resisting LIN driver, as will be explained in Chap. 4. Finally, for EMI frequencies lying above the unity gain frequency of the feedback transistors, the remaining EMI is still filtered by $C_1$, reducing the filter order from a second to a first order. Performing a small signal analysis, the current transfer ratio between the input and the output current is found to

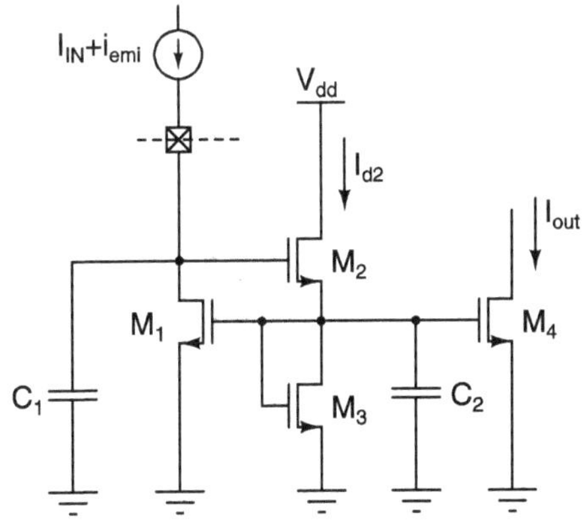


*Figure 3.10.* EMI resisting 4-transistor current mirror.

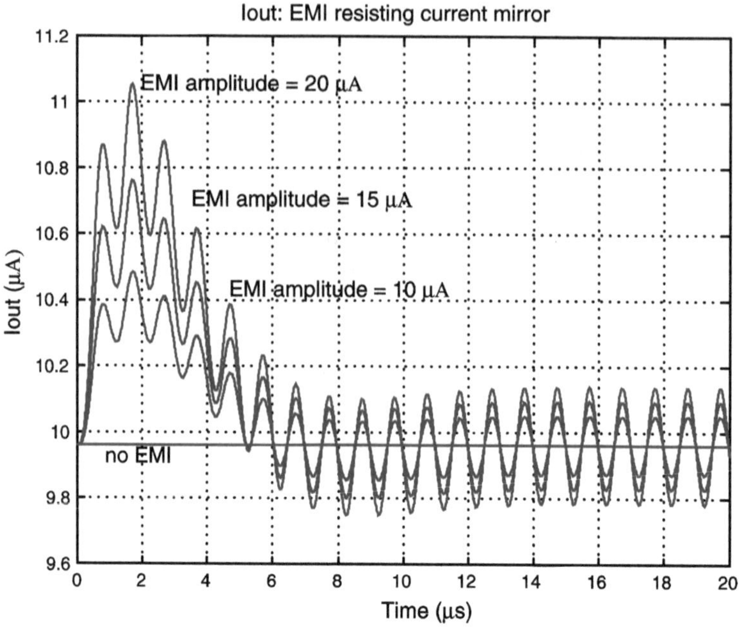


*Figure 3.11.* Output current $I_{out}$ of the improved current mirror: in spite of the applied sinusoidal disturbance at 1 MHz, there is no DC shift.

be equal to:

$$H_4(s) = \frac{i_{out}(s)}{i_{emi}(s)} = \frac{g_{m4}/g_{m1}}{(\frac{C_1 \cdot C_2}{g_{m1} \cdot g_{m2}}) \cdot s^2 + (\frac{C_1 \cdot (g_{m2}+g_{m3})}{g_{m1} \cdot g_{m2}}) \cdot s + 1} \tag{3.39}$$

The output current is then expressed as:

$$I_{out} = \frac{W_2/L_2}{W_1/L_1} \cdot I_{IN} + m \cdot H_4(j\omega) \cdot I_{IN} \cdot \sin(\omega t) \tag{3.40}$$

As specified earlier, capacitor $C_2$ does not cause DC shift, since the gates of $M_1$ and $M_2$ are connected to each other. This is illustrated in Fig. 3.11, where the output current of the improved current mirror is plotted, using the same EMI disturbance and bias current as in the example of the standard current mirror with low-pass $R$-$C$ filter. The size of $M_1$ has been chosen equal to the size of $M_4$ ($W_1/L_1 = W_4/L_4 = 10$ µm/1 µm; $g_{m1} = g_{m4} = 140$ µS) and in the same way $M_2$ has been chosen equal to $M_3$ ($W_2/L_2 = W_3/L_3 = 5$ µm/1 µm; $g_{m2} = g_{m3} = 70$ µS). As a point of comparison, the same arbitrary attenuation of $-40$ dB at 1 MHz has been chosen correspondingly to the previous design examples throughout this section. Capacitors $C_1$ and $C_2$ determine the location of the two poles: these were selected according to a Butterworth filter synthe-

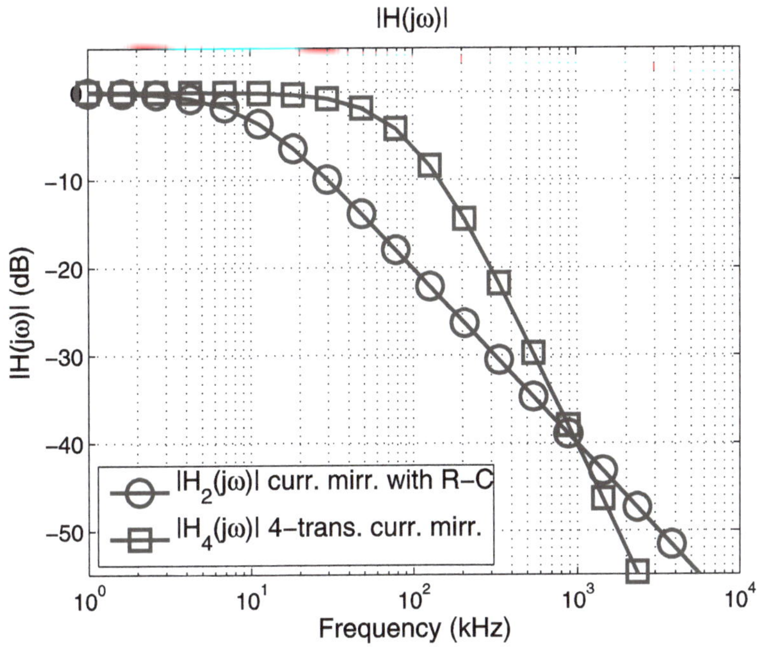


*Figure 3.12.* A comparative AC plot showing the transfer function of the EMI resisting current mirror (Fig. 3.10) using critical damping together with the transfer function of the classic current mirror with a low-pass $R$-$C$ filter in the mirror node (Fig. 3.6c). Both circuits were dimensioned to provide an attenuation of 40 dB at 1 MHz.

sis ($C_1 = 158$ pF, $C_2 = 140$ pF). The motivation behind this choice will be explained in detail in the following paragraph.

As can be seen in Fig. 3.11, the EMI disturbance is strongly attenuated, and after a brief settling, the DC component of $I_{out}$ is identical to the expected value of 10 μA if no disturbance was present. Compared with the transient result of the current mirror with a low-pass filter between its gates (Fig. 3.9), this is a considerable improvement. Figure 3.12 shows the small signal transfer function of the improved 4-transistor current mirror (critical damping case), as well as of the standard mirror with a low-pass filter in the mirror node described in Sect. 5.2. Observe that in both cases, the attenuation at 1 MHz is equal to $-40$ dB. However, using the EMI resisting current mirror, there is no DC shift as can be observed by comparing Fig. 3.11 with Fig. 3.9. The circuit operates is its weakly nonlinear region as long as $m \cdot |H_4(j\omega)| < 1$: paying attention to the fact that $H_4(j\omega)$ uses a second-order filtering, the dynamic range is hereby significantly increased at higher EMI frequencies.

Capacitance is no cheap thing to use in integrated circuits, so it's better to dispose of this resource as economically as possible. Different syntheses can be used to realize a required filter specification. Observing the same attenuation starting from a given EMI frequency while minimizing the sum of $C_1$ and $C_2$ determines the optimal filter synthesis choice. The total capacitance required for this circuit in order to realize an arbitrary low-pass filter function is now derived. To that end, (3.39) is firstly rewritten using the following standard form [Zve67, Dut97]:

$$H(s) = \frac{K \cdot \omega_n^2}{s^2 + 2 \cdot \zeta \cdot \omega_n \cdot s + \omega_n^2} \tag{3.41}$$

where $\omega_n$ is the natural frequency and $\zeta$ represents the damping. Comparing previous expression with the original transfer function given in (3.39), yields the following relationships:

$$\begin{cases} K = \frac{g_{m2} \cdot g_{m4}}{C_1 \cdot C_2} \\ \omega_n = \sqrt{\frac{g_{m1} \cdot g_{m2}}{C_1 \cdot C_2}} \\ \zeta = \frac{g_{m2} + g_{m3}}{2 \cdot \omega_n \cdot C_2} \end{cases} \tag{3.42}$$

Both capacitors $C_1$ and $C_2$ can be expressed as a function of the damping and the natural frequency:

$$\begin{cases} C_1 = \frac{2 \cdot \zeta \cdot g_{m1} \cdot g_{m2}}{\omega_n \cdot (g_{m2} + g_{m3})} \\ C_2 = \frac{g_{m2} + g_{m3}}{2 \cdot \omega_n \cdot \zeta} \end{cases} \tag{3.43}$$

The total capacitance needed is the sum of $C_1$ and $C_2$. If both transistors $M_1$ and $M_2$ are identical (unity gain current transfer), $g_{m2}$ is equal to $g_{m3}$, and the total capacitance is then equal to:

$$C_{tot} = C_1 + C_2 = \frac{2 \cdot \zeta \cdot g_{m1} \cdot g_{m2}}{\omega_n \cdot (g_{m2} + g_{m3})} + \frac{g_{m2} + g_{m3}}{2 \cdot \omega_n \cdot \zeta} = \frac{g_{m1} \cdot \zeta}{\omega_n} + \frac{g_{m2}}{\omega_n \cdot \zeta} \tag{3.44}$$

Previous equation illustrates that the filter synthesis yielding the minimal total capacitance for a fixed cut-off frequency depends on $g_{m1}$, $g_{m2}$, $\omega_n$ and $\zeta$. Different filter syntheses can be used to meet (3.41) while minimizing $C_{tot}$ in (3.44). The three most common have been compared here, namely the critical damping ($\zeta = 1$), Butterworth ($\zeta = 1/\sqrt{2}$) and Chebyshev (3 dB ripple, 0 dB offset and $\zeta = 0.383$) filter syntheses [Zve67]. The normalized filter transfer functions of these three filter syntheses have been plotted in Fig. 3.13. The results are summarized in Table 3.1. In this table, the capacitances $C_1$ and $C_2$ are expressed as a function of the filter $-3$ dB cut-off frequency ($\omega_c$) instead of the natural frequency $\omega_n$. In order to evaluate which filter synthesis yields the smallest total capacitance, ratio $C_{tot}/g_{m1}$ is expressed as a function of $g_{m1}/g_{m2}$, the latter being considered as an independent variable: this approach is summarized in Table 3.2. This way, for a given natural frequency $\omega_n$, the

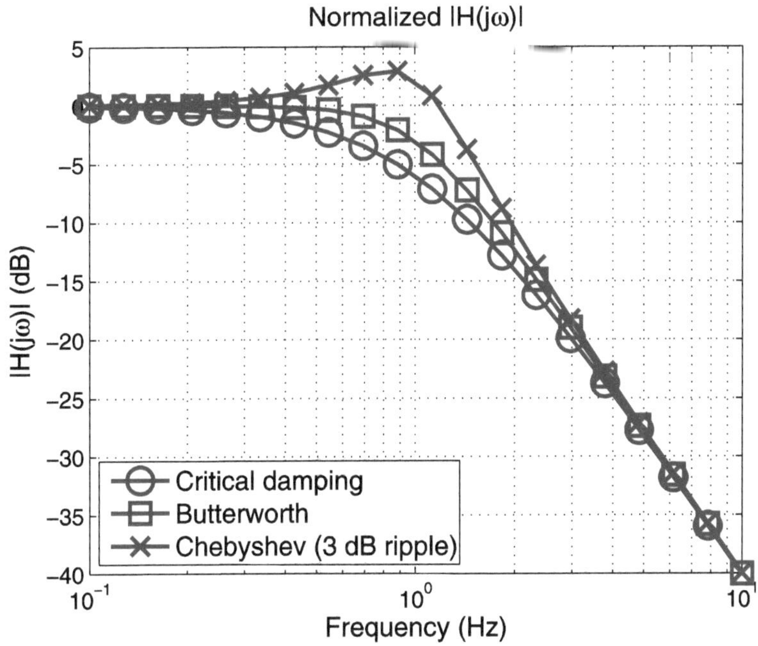


*Figure 3.13.* Normalized second-order filter transfer functions.

*Table 3.1.* Filter syntheses comparison.

| | $\zeta$ | $\omega_n$ | *−3 dB cut-off frequency* $\omega_c$ | $C_1$ | $C_2$ |
|---|---|---|---|---|---|
| Critical damping | 1 | $\sqrt{\frac{g_{m1}\cdot g_{m2}}{C_1\cdot C_2}}$ | $0.639\omega_n$ | $0.639\frac{g_{m1}}{\omega_c}$ | $0.639\frac{g_{m2}}{\omega_c}$ |
| Butterworth | $\frac{1}{\sqrt{2}}$ | $\sqrt{\frac{g_{m1}\cdot g_{m2}}{C_1\cdot C_2}}$ | $\omega_n$ | $\frac{g_{m1}}{\sqrt{2}\cdot\omega_c}$ | $\sqrt{2}\frac{g_{m2}}{\omega_c}$ |
| Chebyshev (3 dB ripple, 0 dB offset) | 0.383 | $\sqrt{\frac{g_{m1}\cdot g_{m2}}{C_1\cdot C_2}}$ | $\omega_n$ | $0.383\frac{g_{m1}}{\omega_c}$ | $2.611\frac{g_{m2}}{\omega_c}$ |

total capacitance of the three filter syntheses can be compared by means of a two dimensional plot.

This is illustrated in Fig. 3.14, where the total needed capacitance divided by $g_{m1}$ is plotted versus the ratio $g_{m1}/g_{m2}$, for a cut-off frequency of 100 kHz. As mentioned previously, $g_{m2}$ has been taken equal to $g_{m3}$. The conclusion of this plot is straightforward: for $g_{m1}/g_{m2} < 1.4$, critical damping yields the smallest total capacitance. When $g_{m1}/g_{m2} > 3.7$, Chebyshev synthesis gives the optimal result. In-between these two values, Butterworth synthesis requires the

*Table 3.2.* Total capacitance per filter synthesis.

| | $C_{tot}$ | $C_{tot}/g_{m1}$ |
|---|---|---|
| Critical damping | $0.639\frac{g_{m1}}{\omega_c} + 0.639\frac{g_{m2}}{\omega_c}$ | $0.639\left(1+\frac{g_{m2}}{g_{m1}}\right)\cdot\frac{1}{\omega_c}$ |
| Butterworth | $\frac{g_{m1}}{\sqrt{2}\cdot\omega_c} + \sqrt{2}\frac{g_{m2}}{\omega_c}$ | $\left(\frac{1}{\sqrt{2}}+\sqrt{2}\frac{g_{m2}}{g_{m1}}\right)\cdot\frac{1}{\omega_c}$ |
| Chebyshev (3 dB ripple, 0 dB offset) | $0.383\frac{g_{m1}}{\omega_c} + 2.611\frac{g_{m2}}{\omega_c}$ | $\left(0.383+2.611\frac{g_{m2}}{g_{m1}}\right)\cdot\frac{1}{\omega_c}$ |

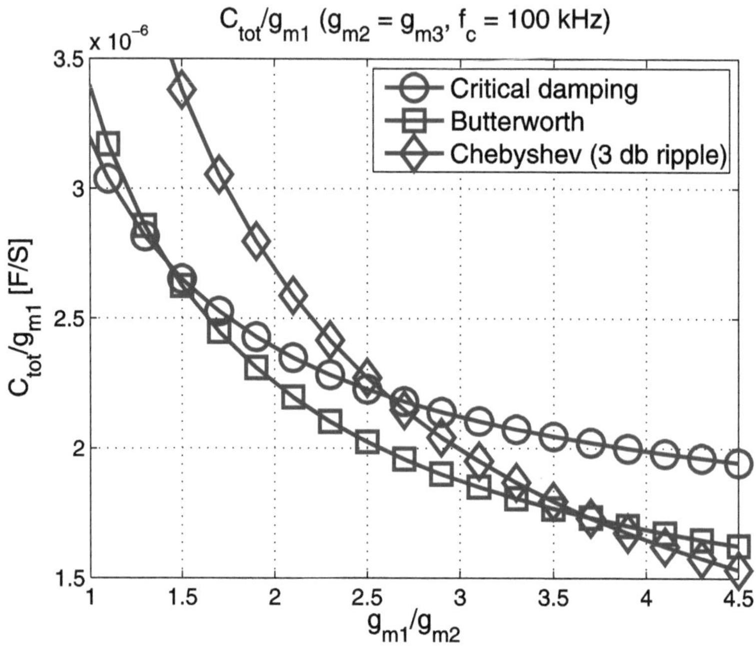


*Figure 3.14.* Comparison of the total required capacitance for the three different filter syntheses, as a function of $g_{m1}/g_{m2}$.

smallest total capacitance. A different insight is provided in Fig. 3.15, namely a plot of the total needed capacitance as a function of the cutoff frequency, for critical damping, Butterworth and Chebyshev syntheses. Observe that there is no point in increasing the ratio $g_{m1}/g_{m2}$ above 3, since the resulting reduction of the total required capacitance $C_{tot}$ becomes quasi-negligible. A similar trend is observed in the critical damping and Chebyshev syntheses.

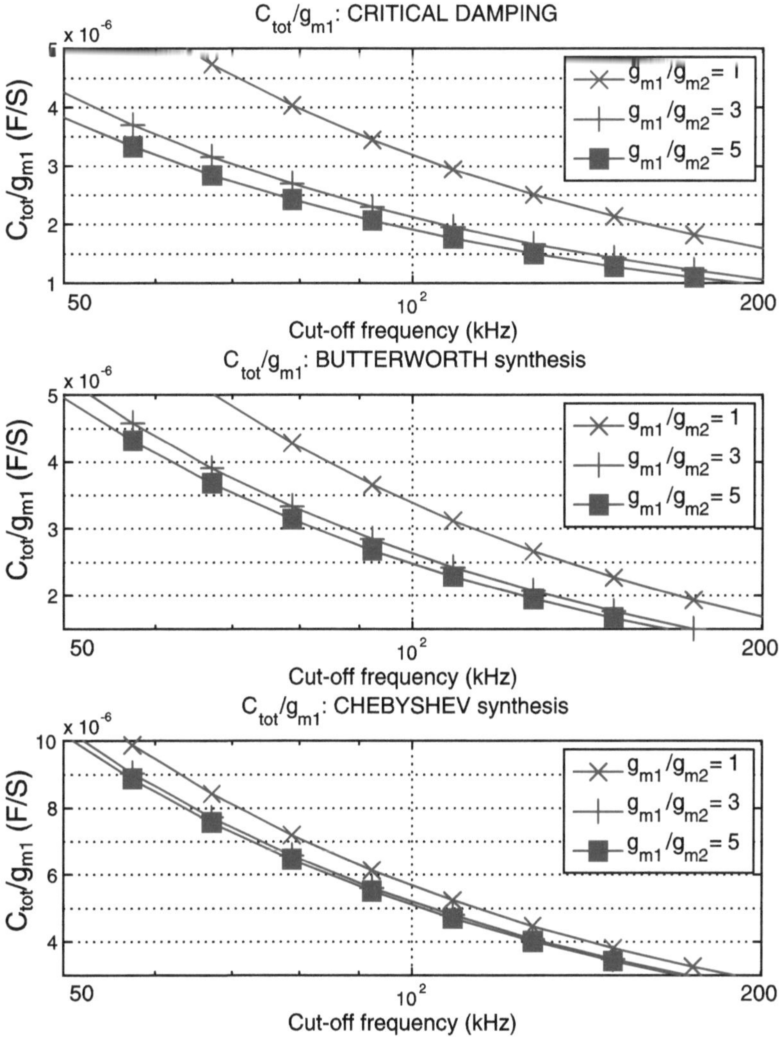


*Figure 3.15.* Comparison of the total required capacitance for the three considered filter syntheses, as a function of the cut-off frequency.

## 5.5 EMI resisting (Wilson totem pole) current mirror

Previous current mirror circuit is improved even more. Refer to the schematic depicted in Fig. 3.10. Removing transistor $M_4$ yields a current mirror of the Wilson type as shown in Fig. 3.16a [Lak94]. The resulting filter function from

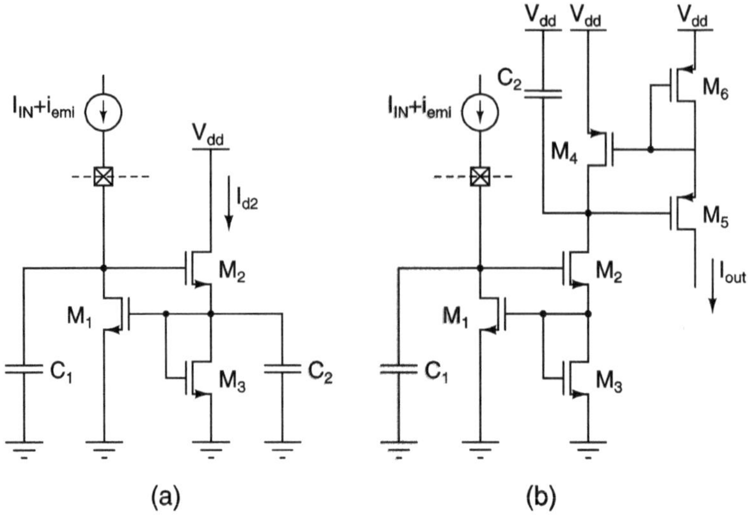


*Figure 3.16.* (a) EMI resisting current mirror without $M_4$ (Wilson current mirror). (b) Wilson totem pole EMI resisting current mirror.

EMI input current to AC drain current of $M_2$ ($i_{d2}$) is equal to:

$$H_5(s) = \frac{i_{d2}(s)}{i_{emi}(s)} = \frac{\frac{g_{m3}}{g_{m1}} \cdot (1 + s \cdot \frac{C_2}{g_{m3}})}{(\frac{C_1 \cdot C_2}{g_{m1} \cdot g_{m2}}) \cdot s^2 + (\frac{C_1 \cdot (g_{m2}+g_{m3})}{g_{m1} \cdot g_{m2}}) \cdot s + 1} \tag{3.45}$$

This transfer function is identical to the previous one, except for the presence of a negative zero. Removing $C_2$ causes this zero to disappear, but at the same time reduces the filter by one order. The current transfer function is then equal to:

$$H_6(s) = \frac{i_{d2}(s)}{i_{emi}(s)} = \frac{\frac{g_{m3}}{g_{m1}}}{(\frac{C_1 \cdot (g_{m2}+g_{m3})}{g_{m1} \cdot g_{m2}}) \cdot s + 1} \tag{3.46}$$

In Fig. 3.16b, two such Wilson mirrors are cascoded (Wilson totem pole). The current transfer function of this Wilson totem pole circuit is expressed as:

$$H_7(s) = \frac{i_{out}(s)}{i_{emi}(s)} = \frac{\frac{g_{m3}}{g_{m1}} \cdot \frac{g_{m6}}{g_{m4}}}{((\frac{g_{m2}+g_{m3}}{g_{m1} \cdot g_{m2}}) \cdot C_1 \cdot s + 1) \cdot ((\frac{g_{m5}+g_{m6}}{g_{m4} \cdot g_{m5}}) \cdot C_2 \cdot s + 1)} \tag{3.47}$$

Assuming out of symmetry that $g_{m2}$ is equal to $g_{m3}$ and that $g_{m5}$ is equal to $g_{m6}$, yields a total capacitance which is equal to:

$$C_{tot} = C_1 + C_2 = \frac{g_{m1} \cdot g_{m2}}{\omega_n \cdot (g_{m2} + g_{m3})} + \frac{g_{m4} \cdot g_{m5}}{\omega_n \cdot (g_{m5} + g_{m6})} = \frac{g_{m1}}{2 \cdot \omega_n} + \frac{g_{m4}}{2 \cdot \omega_n} \tag{3.48}$$

Observe that there is no damping factor $\zeta$ present in (3.48), because both poles are real and lying on top of each other (critical damping case). Comparing (3.48) to (3.44), it can be concluded that, if $g_{m4}$ (in the Wilson totem pole) is made equal to $g_{m2}$ (in the 4-transistor current mirror), the total required capacitance is twice as small for the same natural frequency $\omega_n$, in case of critical damping and using an identical input transistor $M_1$. Ensuring that $g_{m4}$ is smaller than $g_{m1}$ yields an even smaller total capacitance. Note that the immunity against charge pumping is preserved.

## 5.6 Comparison of EMI susceptibility of current mirrors

The EMI susceptibility of various current mirrors has been examined in previous sections: a synopsis is presented in Table. 3.3. Figure 3.17 compares the total capacitance per units of $g_{m1}$ needed in the four-transistor mirror and in the Wilson totem pole for realizing an arbitrary attenuation of 40 dB at an EMI frequency of 1 MHz, against $g_{m1}/g_{m2}$ ($g_{m1}/g_{m4}$ for the Wilson totem pole). It is clearly seen in this figure that a respectable capacitance reduction is achieved in the Wilson totem pole circuit. Observe that there are three main disadvantages of using the Wilson totem pole mirror compared to the 4-transistor EMI resisting current mirror:

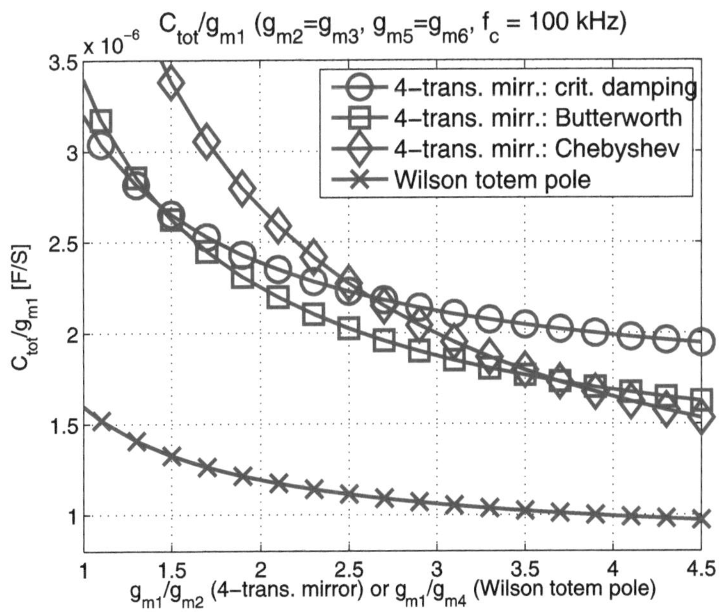


*Figure 3.17.* Comparison of the total required capacitance. Observe that the Wilson totem pole current mirror yields the smallest total capacitance for the same transistor sizes.

*Table 3.3.* Current mirror comparison in terms of EMI susceptibility, input dynamic range and minimum required supply voltage.

| *Current mirror* | *Figure* | *DC shift* | *Input dynamic range* | *Required supply voltage* |
|---|---|---|---|---|
| Classic current mirror … | Fig. 3.6a | no | $\frac{g_{m2}}{g_{m1}} \cdot m < 1$ | $1 \cdot V_{GS}$ |
| … with capacitor in mirror node | Fig. 3.6b | no | $\frac{g_{m2}}{g_{m1}} \cdot \frac{m}{1+\frac{s \cdot C}{g_{m1}}} < 1$ | $1 \cdot V_{GS}$ |
| … with a low-pass $R$-$C$ in mirror node | Fig. 3.6c | yes | $\frac{g_{m2}}{g_{m1}} \cdot \frac{m}{1+s \cdot R \cdot C} < 1$ | $1 \cdot V_{GS}$ |
| … with a low-pass $R$-$C$ in drain | Fig. 3.6d | no | $\frac{g_{m2}}{g_{m1}} \cdot \frac{m}{s \cdot (C \cdot R+\frac{C}{g_{m1}})+1} < 1$ | $1 \cdot V_{GS} + I_{IN} \cdot R$ |
| 4-transistor | Fig. 3.10 | no | $\frac{g_{m4}}{g_{m1}} \cdot \frac{m}{\left(\frac{C1 \cdot C2}{g_{m1} \cdot g_{m2}}\right) \cdot s^2+\left(\frac{C1 \cdot (g_{m2}+g_{m3})}{g_{m1} \cdot g_{m2}}\right) \cdot s+1} < 1$ | $2 \cdot V_{GS}$ |
| Wilson totem pole | Fig. 3.16b | no | $\frac{g_{m3}}{g_{m1}} \cdot \frac{m}{\left(\frac{C1 \cdot (g_{m2}+g_{m3})}{g_{m1} \cdot g_{m2}}\right) \cdot s+1} < 1$ | $3 \cdot V_{GS} + V_{DSsat}$ |

*Table 3.4.* Design values.

| | *Transistor sizes* | $I_{IN}$ *(DC)* |
|---|---|---|
| 4-transistor current mirror (critical damping) | $W_1/L_1 = W_4/L_4 = 10$ µm/1 µm<br>$W_2/L_2 = W_3/L_3 = 5$ µm/1 µm<br>$g_{m1} = 140$ µm, $g_{m2} = 70$ µm | 10 µA |
| 4-transistor current mirror (Butterworth) | $W_1/L_1 = W_4/L_4 = 10$ µm/1 µm<br>$W_2/L_2 = W_3/L_3 = 5$ µm/1 µm<br>$g_{m1} = 140$ µm, $g_{m2} = 70$ µm | 10 µA |
| 4-transistor current mirror (Chebyshev) | $W_1/L_1 = W_2/L_2 = 10$ µm/1 µm<br>$W_2/L_2 = W_3/L_3 = 5$ µm/1 µm<br>$g_{m1} = 140$ µm, $g_{m2} = 70$ µm | 10 µA |
| Wilson totem pole current mirror | $W_1/L_1 = W_5/L_5 = W_6/L_6 = 10$ µm/1 µm<br>$W_2/L_2 = W_3/L_3 = W_4/L_4 = 5$ µm/1 µm<br>$g_{m1} = 140$ µm, $g_{m4} = 70$ µm | 10 µA |

- **Smaller dynamic range:** the EMI ripple at the input node is larger, and this means that heavy nonlinear distortions appear more quickly. This is because the input dynamic range is increased by $1/|H_6(j\omega)|$, which is a one pole transfer function instead of $1/|H_4(j\omega)|$, which is a two pole transfer function.

- **Orientation:** in order to achieve a second-order filtering, two Wilson totem pole current mirrors must be cascaded. This equally means that a sinking current source output is transformed into a sourcing current source output and vice versa.

- **Higher supply voltage:** The 4-transistor EMI resisting current mirror requires at least two $V_{GS}$ voltages at its input node in order to function correctly. The Wilson totem pole current mirror requires at least $3V_{GS} + V_{DSsat}$ in order to keep $M_2$, $M_3$ and $M_4$ in saturation, which is a considerable requirement to meet in present day integrated technologies.

*Example*: Both EMI resisting current mirrors depicted in Figs. 3.10 and 3.16b using different filter syntheses were compared using a concrete design example which was simulated in a standard CMOS 0.35 µm technology. Both circuits were designed to provide an attenuation of 40 dB starting from 1 MHz. The design values are enumerated in Table 3.4, while Table 3.5 lists the total required capacitor values. Not surprisingly, the Wilson totem pole current mirror requires the smallest total capacitance, which is not surprising considering the fact that it requires half of the total capacitance of a four-transistor current mirror designed for critical damping. While the capacitance values are by no

*Table 3.5.* Total required capacitor values.

| | $C_1$ *(pF)* | $C_1$ *(pF)* | $C_{tot}$ *(pF)* |
|---|---|---|---|
| 4-transistor current mirror (critical damping) | 223 | 111 | 334 |
| 4-transistor current mirror (Butterworth) | 158 | 158 | 316 |
| 4-transistor current mirror (Chebyshev) | 85 | 291 | 376 |
| Wilson totem pole current mirror | 111 | 55 | 167 |

means small, they are still quite integrable without having to resort to excessive areas. Even more importantly, both EMI resisting current mirror topologies present a high immunity against EMI induced DC shift.

## 6 Case study 4: EMI susceptibility in ESD protections

Triboelectric, ionic, inductive and direct charging generate electrostatically charged objects: the voltage in-between them typically rises several ten thousands volts in everyday life, depending on the used materials as well as on the charging conditions [Gla07]. When referring to integrated circuits, electrostatic discharge (ESD) occurs when a charged external object is connected to one of the IC external connections. Because the internal capacitances present at each input or output are typically very small, this charge rebalancing produces a large voltage spike, hereby possibly damaging the IC depending on the gate oxide voltage breakdown level. Moreover, the inflicted damage increases drastically in newer technologies using thinner gate oxides and smaller channel lengths [Ben06]. Therefore, it is hardly surprising that ESD is one of the most common causes of failures appearing in present day silicon IC's: in 10% up to 25% of the cases, ESD is identified as being the primary reason of failure in IC's, according to [Cro03]. All these facts fuel the ever increasing need and awareness for robust and reliable ESD protection circuits – a tendency quite similar to the ever growing interest in EMI robust IC's.

In order to alleviate ESD-stress, ESD protective devices are used to limit the voltage spikes which are applied to the circuit [Raz01]. Consequently, in most cases, ESD protections contain one or more nonlinear devices which clamp the voltage on an I/O node between predefined levels. Mainly for this reason, these protective structures need to be taken into account when considering the

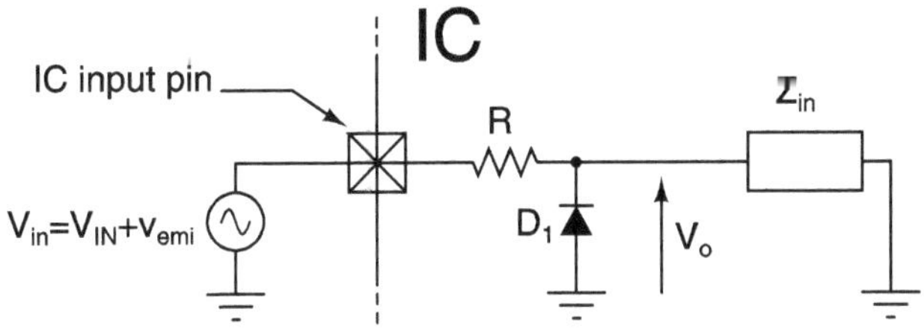


*Figure 3.18.* A typical nonlinear ESD protection.

complete EMC behavior of a given IC [Loe07]. In addition, when designed without attention for EMC, they may impair and even completely destroy the full chip operation owing to EMI which is injected into the I/O terminals. For the above reasons, the basic EMC considerations concerning ESD protections are briefly discussed is this paragraph. Consider the simplified ESD protection depicted in Fig. 3.18 which clips negative voltage spikes. Very often, a second diode connected to the positive power supply is added as well, in order to limit the ESD overshoots, yielding the well-known double diode ESD protection (Fig. 3.23a). However, in present case study, the upper diode is left out in order to simplify the calculations. The voltage $V_{in}$ which is applied at the input of this ESD protection, consists of a DC bias value ($V_{IN}$) and an AC component ($v_{emi}$), which represents a sinusoidal EMI disturbance that is superimposed on the IC input pin in question:

$$v_{emi} = \hat{v} \cdot \sin(\omega t) \tag{3.49}$$

$Z_{in}$ symbolizes the input impedance of the IC seen after the ESD protection: however, it will be assumed that $Z_{in}$ does not interfere with the operation of the ESD protection (in other words, $Z_{in}$ is infinitely large). The behavior of this simplified ESD protection is analyzed in order to study and explain how EMI affects the subsequent circuit, represented by input impedance $Z_{in}$ [Red07c]. Once again, the weak nonlinear behavior must be distinguished from the strong nonlinear behavior: the former is discussed first.

## 6.1 Weak nonlinear distortion in ESD protections

As long as the input signal is small ($V_{IN} > 0, |\hat{v}| < V_{IN}$), the ESD diode depicted in Fig. 3.18 ($D_1$) is reversely biased. Except for a negligible saturation current $I_S$, no DC current flows through this diode, and consequently, the diode is modeled purely like a nonlinear junction capacitor. In practice, resistor $R$ may contain nonlinear components which impact the EMI behavior of the ESD protection as well. Both nonlinearities and their impact on the circuit are evaluated in the following two subsections. In order to check their respective effects separately, it will be assumed at first that $R$ is perfectly linear, and that the only nonlinearity in this circuit stems from the nonlinear junction capacitance of the diode. After this, the effect of a nonlinear resistor $R$ is derived in detail.

- **EMI amplitude is small: $R$ is linear, $C$ is nonlinear:** Assuming at first that resistor $R$ is linear, the only nonlinear element present in this simplified circuit is the nonlinear junction capacitance of the ESD diode, which is expressed as follows [San06]:

$$C_j = \frac{C_{j0}}{\sqrt{1 - \frac{V_D}{\psi_0}}} \tag{3.50}$$

where $C_{j0}$ represents the zero-bias junction capacitance, $\psi_0$ the junction built-in potential and $V_D$ the voltage across the junction (in this particular case, $V_D = -V_o$). However, as indicated in Sect. 5.1, a linear capacitor does not cause charge pumping, since it does not conduct a net current over time. The same reasoning applies to a nonlinear capacitor: although a nonlinear AC current is flowing through it, its average value is equal to zero. Consequently, the mathematical derivation performed in Sect. 5.1, (3.28)–(3.30), can be extended to nonlinear capacitors. This remarkable observation is now verified in the basic ESD protection circuit, depicted in Fig. 3.18. Solving $V_o(t) = f(V_{in}(t))$, and substituting (3.50), yields the following nonlinear differential equation:

$$\frac{d}{dt}\left(\frac{V_o(t)}{\sqrt{1 + \frac{V_o(t)}{\psi_0}}}\right) = \frac{V_{IN} + \hat{v} \cdot \sin(\omega \cdot t) - V_o(t)}{R \cdot C_{j0}} \tag{3.51}$$

Unfortunately, this differential equation is extremely cumbersome to be solved by hand, and it is therefore evaluated using a numeric example. To this end, the nonlinear junction capacitance $C_j$ is expressed as a power series as a function of the voltage across the capacitor ($V_o$):

$$C_j = C_{j0} \cdot (1 + c_1.V_{res} + c_2 \cdot V_{res}^2 + \cdots) \tag{3.52}$$

where $c_1, c_2, \ldots$ are the coefficients of the nonlinear capacitor power function. The strength and the type of this nonlinearity depends on the value of coefficients $c_i$: out of simplicity, these coefficients were chosen equal to one, and the nonlinear terms have been limited to the square power of $V_o$. This yields the following expression:

$$C_j = C_{j0} \cdot (1 + V_o + V_o^2) \tag{3.53}$$

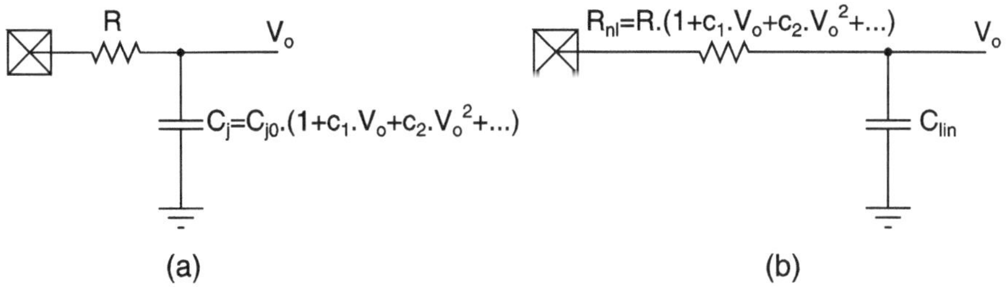


*Figure 3.19.* (a) Simplified circuit with a nonlinear capacitor $C_j$. (b) Simplified circuit with a nonlinear resistor $R_{nl}$.

The resulting schematic is depicted in Fig. 3.19a. $R$ and $C$ have been dimensioned to yield a cut-off frequency of 10 kHz. Circuit simulations comparing the output waveform using respectively a linear and a nonlinear capacitor for an EMI frequency lying above as well as below the filter cut-off frequency, are depicted in Fig. 3.20. The results show that the mean value of $V_o(t)$ over time is equal to $V_{IN}$, meaning that the presence of a nonlinear capacitor does not contribute to DC shift, as was rightly predicted at the beginning of this paragraph. Returning to the studied ESD protection circuit, this signifies that the nonlinear junction capacitor $C_j$ has no influence on the mean value of $V_o(t)$, and may therefore be replaced by a linear equivalent ($C_{lin}$) as far as the calculation of the mean value of $V_o(t)$ is concerned. Using (3.51), combined with the boundary value condition that the mean value of $V_o(t)$ is equal to $V_{IN}$ at time zero, yields:

$$\overline{V_o(t)} = \overline{\frac{V_{IN} \cdot (1 + \omega^2 \cdot R^2 \cdot C_{lin}^2) - \hat{v} \cdot \omega \cdot R \cdot C_{lin} \cdot \cos(\omega \cdot t) + \hat{v} \cdot \sin(\omega \cdot t)}{1 + \omega^2 \cdot R^2 \cdot C_{lin}^2}}$$
$$= V_{IN}$$

Previous expression clearly illustrates once more that a capacitor (be it linear or nonlinear) does not introduce DC shift: it is solely responsible for accumulating a rectified signal, hereby generating DC shift. This striking and fundamental property contains major repercussions when designing EMI resisting circuits.

- **EMI amplitude is small: *R* is nonlinear, *C* is linear:** Next, the impact of a nonlinear resistor $R_{nl}$ is evaluated. The nonlinearity of a resistor is expressed as a power function:

$$R_{nl} = R \cdot (1 + c_1 \cdot V_{res} + c_2 \cdot V_{res}^2 + \cdots) \tag{3.54}$$

where $V_{res}$ is the voltage across the resistor ($V_{res} = V_{in} - V_o$), and $c_1$, $c_2$, ... are the coefficients of the nonlinear resistor power function. Since

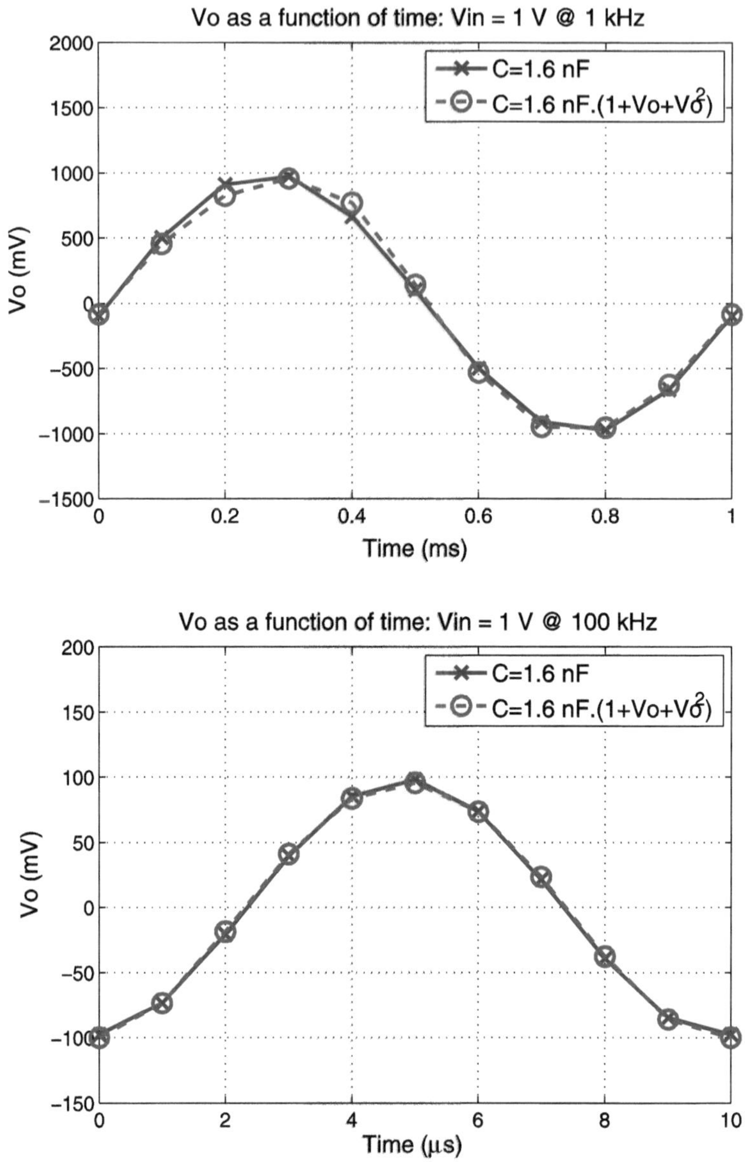


*Figure 3.20.* Effect of a nonlinear capacitor $C_j$. Observe that there is no DC shift.

it has previously be shown that a nonlinear capacitor causes no DC shift, $C_j$ is substituted by an equivalent linear capacitor ($C_{lin}$). The resulting small signal equivalent is depicted in Fig. 3.19b. Solving $V_o(t) = f(V_{in}(t))$

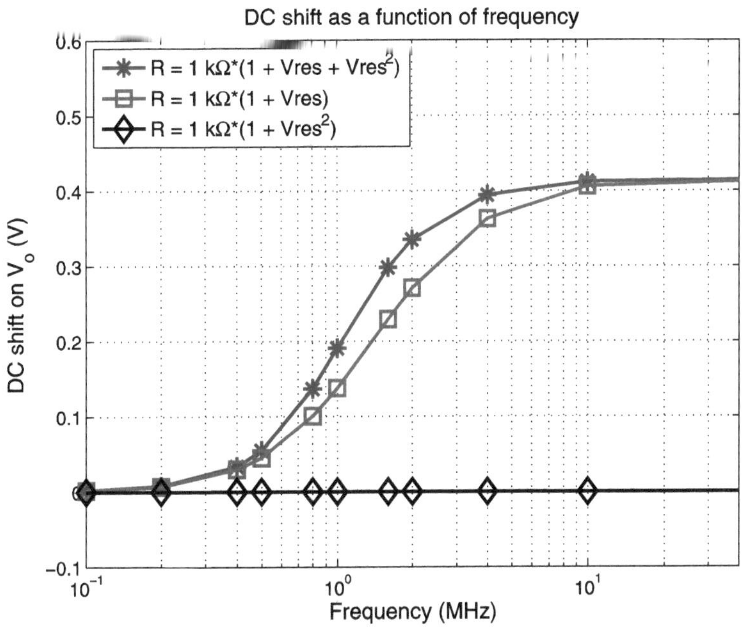


*Figure 3.21.* Effect of a nonlinear resistor $R$. The resistor nonlinearity power function coefficients have been made equal to 1: as a consequence, the curves reflect the effect of even and odd nonlinearity without concentrating on absolute DC shift values.

yields the following nonlinear differential equation:

$$\frac{dV_o(t)}{dt} = \frac{1}{R_{nl} \cdot C_{lin}} \cdot (V_{IN} + \hat{v} \cdot \sin(\omega \cdot t) - V_o(t)) \tag{3.55}$$

where $R_{nl}$ is expressed as in (3.54). This expression was solved numerically as previously, and the DC shift of $V_o$ has been plotted in Fig. 3.21 for even and odd resistor nonlinearities. In this plot, $R$ is equal to 1 kΩ, and capacitor $C_{lin}$ is equal to 100 pF. The EMI amplitude is equal to 1 V. Observe that well below the cut-off frequency of the filter, there is no DC shift, because almost no AC current is flowing through the capacitor, meaning that there is no accumulation. In the same way, above the filter cut-off frequency, a non-linear current is flowing through the network, and the capacitor accumulates the rectified signal value. When the resistor nonlinearity is odd (cubic), the rectification is symmetrical, and so there is no DC shift: on the other hand, DC shift occurs when the resistor nonlinearity is even (quadratic), because the rectification is asymmetric. This corroborates the general observations that were made at the beginning of this chapter. The difference between the first and the second plot in Fig. 3.21 is caused by the shifted cut-off frequency.

## 6.2 Strong nonlinear distortion in ESD protections

When the input voltage drops below the clamping voltage of the diode in the basic ESD protection circuit depicted in Fig. 3.18, it briefly turns on and consequently clips the output voltage $V_o$. Using the diode equation, the output signal waveform can be expressed as follows [San06]:

$$V_{in}(t) = V_{IN} + \hat{v} \cdot \sin(\omega \cdot t) = V_o(t) - R \cdot I_S \cdot e^{\frac{-V_O}{n \cdot V_T}} \quad (3.56)$$

where $I_S$ represents the saturation current and $V_T$ is the thermal voltage, which equals $kT/q$ and is approximately 26 mV at room temperature. The upper plot depicted in Fig. 3.22 shows the effect of clipping on the EMI signal: observe that since clipping is nonlinear and since it takes place asymmetrically, the mean DC voltage is no longer equal to $V_{IN}$. This seriously alters and degrades the performance of the circuits connected to that input pin, since they are no longer correctly biased. The bottom plot in Fig. 3.22 shows the resulting DC shift as a function of the EMI amplitude induced in the circuit depicted in Fig. 3.18, for different input DC bias voltages: the higher $V_{IN}$, the smaller the clipping and the corresponding DC shift.

Based on the previous observations, this DC shift can be reduced or avoided altogether by clipping the high input signal in a symmetric way. Very often, ESD protective circuits are designed as depicted in Fig. 3.23. In this well-known double diode structure, the diodes clamp the injected signal between ground and the positive power supply voltage, which causes the large EMI to clamp asymmetrically depending on the bias voltage $V_{IN}$, hereby generating rectification as shown previously. A possible improvement for this structure has been presented in [Mey03], and is depicted in Fig. 3.24. This circuit consists of two clamping structures triggering in a symmetrical way. The clamping voltage $V_{clamp}$ lies typically at a higher value (e.g. 12 V), meaning that firstly, the immunity level is increased well above the supply voltage, and secondly, that in case the clamping voltage is exceeded by the EMI signal, it is limited in a symmetrical way, hereby preventing DC shift. The latter is clearly only true as long as the bias voltage on this node ($V_{IN}$) is equal to zero.

Another possible solution is to use passive linear networks in order to protect the IC from ESD-stress, instead of clamping the input signals using highly nonlinear devices. These techniques have been examined and successfully applied in the design of integrated RF CMOS low noise amplifiers (LNA's) in [Ler01, Ler03] and [Ler05]. The concept behind these passive ESD protective linear networks is straightforward, and is depicted in Fig. 3.25. An internal inductor, $L_{esd}$, is added in order to tune out the input parasitic capacitance ($C_{par}$) at the operation frequency ($L_b$ is the parasitic series inductance). The tuned tank circuit presents a high impedance for the high frequency input signal, while filtering the low frequency ESD pulse. Indeed, ESD signals are rela-

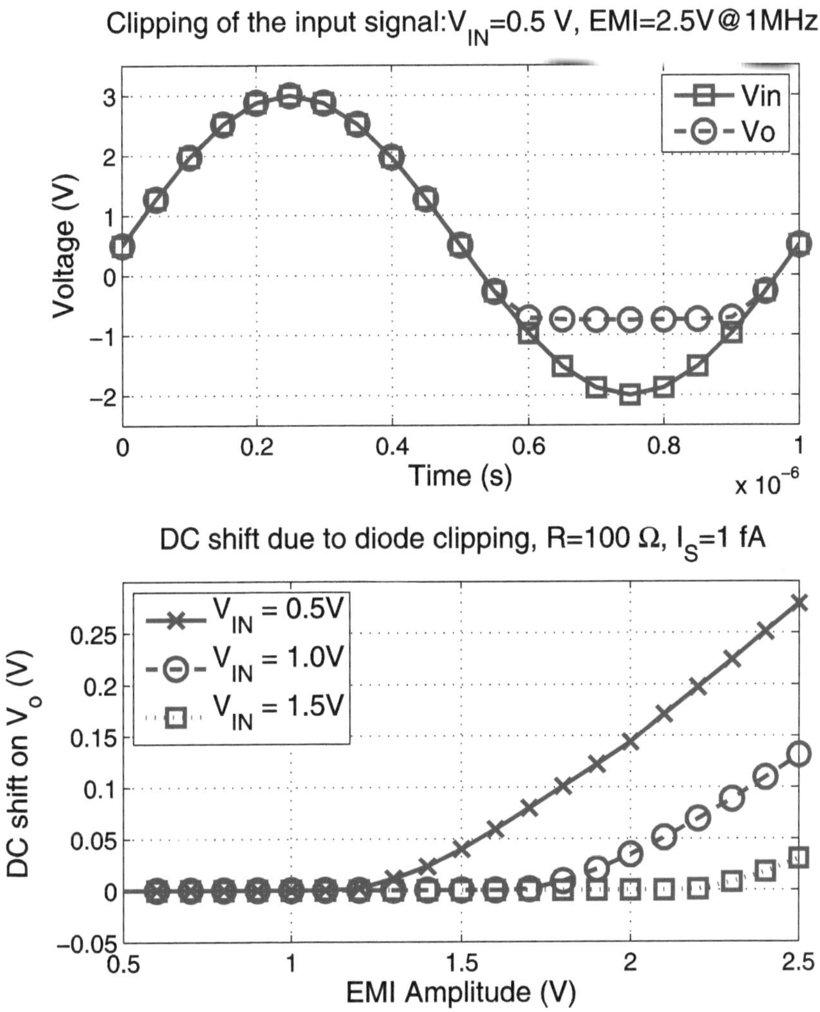


*Figure 3.22.* Large signal behavior of a typical nonlinear ESD protection.

tively low frequent compared to the RF input signal: depending on the used model, their frequency spectrum rolls off in the low MHz range using the human body model (HBM), while resonating up to 500 MHz, using the charged device model (CDM). Cancellation techniques improving the basic passive network protection depicted in Fig. 3.25 have been presented in order to deal with high frequency oscillations resulting from the latter [Hyv03]. The advantage of using these from an EMC point of view is twofold. First of all, these passive networks comprised of integrated inductors and capacitors do not clamp the input voltage, meaning that they do not cause rectification. Secondly, they consist

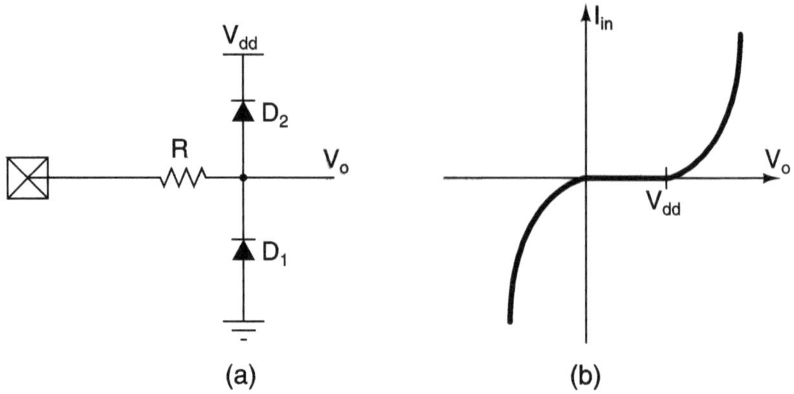


*Figure 3.23.* (a) Typical asymmetrical double diode ESD protection. (b) $I$ to $V$ characteristic of the double diode ESD protection.

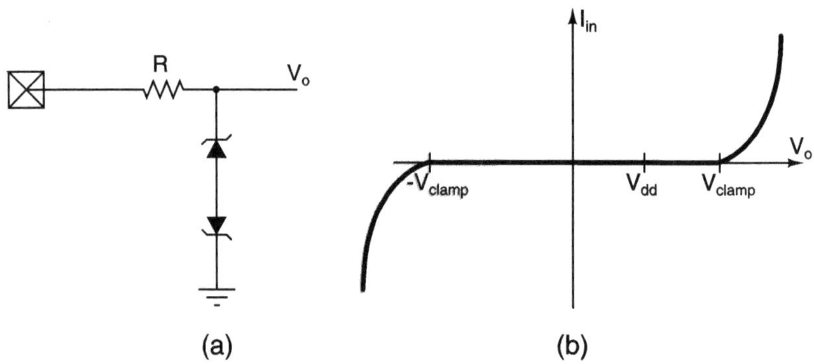


*Figure 3.24.* (a) ESD protection using two clamping devices. (b) $I$ to $V$ characteristic of the ESD protection using two clamping devices.

of integrated capacitors and inductors, whose nonlinearities do not contribute to DC shift, as has been illustrated anteriorly (obviously, the benefic properties ascribed to capacitors in the previous paragraph, apply to inductors as well). However, in the present realizations, these linear ESD protections are only useable for high frequency and narrowband input signals [Ler03].

## 6.3 ESD protections: general conclusions

Refer once more to the basic ESD protection depicted in Fig. 3.18. For small EMI amplitudes, and considering that resistor $R$ is linear, $V_o$ becomes nonlinear owing to the diode junction capacitance: however, since this nonlinear junction capacitance does not conduct a mean net current, there is no DC shift and the mean value of $V_o(t)$ is equal to $V_{IN}$. If resistor $R$ is nonlinear, there is a net DC shift at higher frequencies if this nonlinearity contains even com-

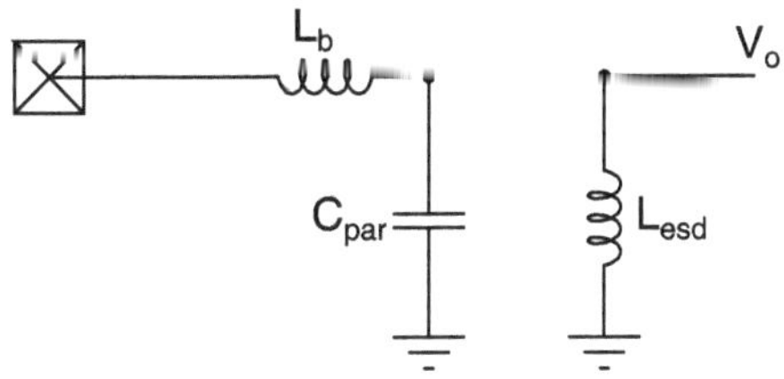


*Figure 3.25.* Passive ESD protection.

ponents. This conclusion is tremendously important when designing suitable ESD protections on EMI sensitive pins. For large EMI amplitudes which swing beyond the clamping voltage of the ESD protection, the resulting $V_o$ becomes heavily nonlinear. Theoretically, provided that this nonlinearity is symmetrical around the DC bias voltage, no rectification occurs: however, such compensation techniques are not very accurate, and are therefore difficult to realize effectively in practice. Another possibility is to increase the clamping voltage in a symmetrical way, in order to reduce or eradicate the heavy nonlinearity, but this is heavily technology dependent and must be checked case by case, since this equally increases the circuit's susceptibility to ESD pulses. Another possibility is to use ESD protections which consist of linear elements as in [Ler03], since these do not distort the EMI signal nonlinearly and consequently generate no DC shift.

# 7 EMI induced DC shift

Collecting the information in this chapter, and considering the four discussed case studies, it is clear that EMI induced DC shift is the worst appearing EMC phenomenon at integrated circuit level, disrupting the sound operation of IC's, and sometimes even debiasing them completely. It is therefore of paramount importance to identify the parameters which induce DC shift. As has been studied in this chapter, the latter is generated by the accumulation of asymmetrically rectified signals, which are in turn produced by resistors whose nonlinearity is even-ordered. In principle, getting rid of nonlinear resistors (e.g. by applying linearization) ensures a DC shift-free circuit operation. This is, however, easier said than done. As discussed anteriorly, preventing accumulation by increasing the bandwidth of the circuit, decreases DC shift. However, since the bandwidth is typically limited, DC shift can always occur in practice, depending on the interfering frequencies: consequently, filtering high frequency EMI disturbances before they reach sensitive circuit nodes remains mandatory. Finally, in order to simplify the used terminology in this work, the accumulation process will be implicitly assumed when talking about DC shift [2].

[2] An example of how complicate obvious descriptions would otherwise sound like, is offered in the footnote on page 51.

# Chapter 4

# EMI Resisting Analog Output Circuits

*Electromagnetic effects can cause impressive disasters that urge us to control the problem. One example is the catastrophe with H.M.S. Sheffield during the Falkland crisis. An Exocet missile hit this frigate because its search radar was switched off. It was switched off because it was known that the satellite communication system was interfered with by this radar. At the time of the disaster some officers used this communication link to talk with their prime minister. (...) The 'disaster philosophy' is already known to many EMC engineers who every now and then make use of a disaster to get new budgets.*

—Quoted from [Gro02]

## 1 Introduction

There are numerous integrated output circuits, and an attempt to study all of them at the same time in view of understanding and improving their susceptibility to EMI would quickly prove to be a futile effort. However, under the assumption that similar output structures experience the same type of EMC issues, it is not too farfetched to claim that the same solutions which can be applied to one type of circuit may improve the EMC behavior of a related topology. After all, as was indicated in Chap. 3, although EMC problems originate almost in any type circuit – basically, they are spawned by the accumulation of a nonlinear signal – their effect is usually limited to signal distortion and DC shift: the latter is particularly harmful since it tinkers with the circuit's operating point, alters its bias currents and depending on the strength of the interfering signal even possibly debiases the full circuit completely.

To this end, this chapter classifies and characterizes the electromagnetic susceptibility of CMOS analog integrated output structures in a general way. Applying the observations established in Chap. 3 on more complex (output) circuits, the originating EMC problems are identified using small signal approximations in order to detect the appearing EMC effects. The advantage of

J.-M. Redouté, M. Steyaert, *EMC of Analog Integrated Circuits,*
Analog Circuits and Signal Processing,
DOI 10.1007/978-90-481-3230-0_4, © Springer Science+Business Media B.V. 2010

using small signal approximations to do so resides in the fact that the mathematical complexity is greatly reduced, since a purely linear analysis is required to detect the appearing EMI issues [Red07b]. However, as has been illustrated anteriorly, small signal analyses are not sufficient by themselves to quantize and appreciate the full impact of EMI on a given circuit. Therefore, the output circuit characterizations are equally conducted from a large signal perspective, in order to check at which point the circuit is interfered with by strong nonlinear distortions using anhysteretic power series. This approach is valid in the event that the power series converges, which is the case as long as the active devices are not pulled out of their operating region by large EMI signals [Wam98]. Taking the previous comments into account, it is shown further on that CMOS analog integrated output structures can be distinctively grouped into two major categories, namely the output circuits driving the load at the transistor source (common-drain type of circuits), and those driving the load at the transistor drain (common-source type of circuits). The major properties of both circuit types are derived mathematically, and compared to each other. As illustrated, common-source circuits exhibit a much higher immunity to EMI which is injected into their output terminals compared to their common-drain counterparts. After developing this theoretical foundation, two case study examples are studied and evaluated in detail, namely:

- **Case study 1: an EMI resisting integrated DC current regulator using an external trimming resistor.** This case study concentrates on improving the EMC behavior of an externally trimmed DC current regulator. Being originally designed as a common-drain output type, the appearing EMI problems are identified and matched to the theoretical observations. An EMI resisting current regulator structure with a rerouted feedback loop using a common-source output type is introduced and described: the resulting high immunity to EMI of this improved design is confirmed with measurements of a test-IC.

- **Case study 2: an EMI resisting integrated Local Interconnect Network (LIN) driver.** Although the classic LIN driver design is a common-source output type, a high capacitive coupling is essentially responsible for a low EMI immunity. As a consequence, the EMI disturbances injected in the output of this circuit easily propagate to the gate of the output driving transistor, hereby irrevocably distorting the output signal. Using the general derived observations in order to improve the robustness of the original design, a much higher immunity to EMI can be achieved. Two test-IC's are presented in this case study. The first test-IC shows how the appearing EMC problems are solved using a modified design concept, however, at the cost of a large and continuous power consumption as well as a mediocre behavior of the output signal slopes. Both issues are dealt with and solved in the second EMI resisting LIN driver topology, by firstly adding a smart-power capability to the circuit, namely an extra feature to detect if EMI is present and adjust the current consumption accordingly, and secondly by redesigning the circuit blocks which are responsible for the non-negligible slope pumping. The superior performance of the second topology is again verified with measurements. For both test-IC's, measurements confirming the theoretical deductions are presented.

# 2 Categorization of analog output structures

## 2.1 Common-drain output circuits

Output circuits which are required to drive a low impedance load with a negligible loss of signal level or gain are very often common-drain structures. Consider as an example the well-known source follower (Fig. 4.1a) as well as the class B complementary "push-pull" output stage (Fig. 4.1b) [Gra01]. In both cases, the source(s) of the driving transistor(s) is (are) connected to the output, ensuring that the output node is driven with a low impedance. Additionally, both circuits are used in combination with a load, which in an integrated environment is sometimes realized as a high impedance current source. However, for the purpose of generality, an explicit load resistor $R_L$ is considered here. This section examines the effect of EMI coupling into the output node of this type of circuits [Red07b]. To this end, an EMI AC voltage source ($v_{emi}$) is connected to the output of a source follower circuit by means of a coupling capacitor $C_c$, as in Fig. 4.1c. This capacitor represents a perfect AC coupling between the EMI source and the circuit in question, and is by default equal to 6.8 nF, as specified in the DPI specification [DPI]. The purpose of $C_c$ is to couple ideally the EMI signal into the circuit without interfering with the circuit behavior, and therefore it will be shorted in the following small signal calculations. A forward EMI power is injected through this coupling capacitor

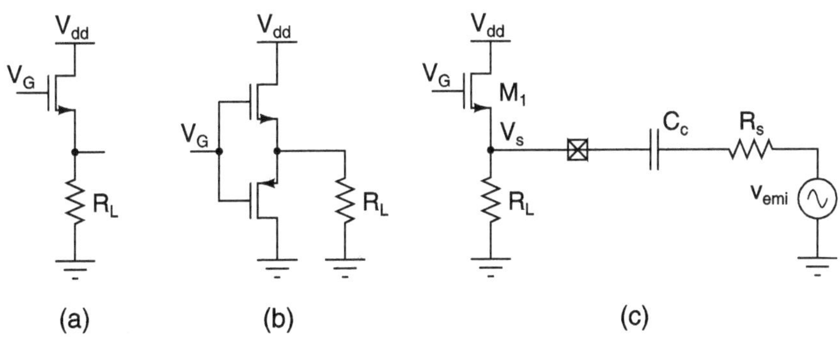


*Figure 4.1.* (a) Source follower. (b) Push-pull class B output stage. (c) Source follower with EMI injection at its output.

into the circuit. The EMI voltage source itself contains an internal impedance which is modeled as source resistance $R_s$. The nominal value of this source resistance is equal to 50 $\Omega$, but can be increased up to 150 $\Omega$, according to the DPI specification, if the protection resistance $R_p$ is taken into account [1]. The DPI measurement setup has been covered in Chap. 2.

Clearly, the EMI varies the gate-source voltage of transistor $M_1$, and consequently alters the average drain current of $M_1$. The average drain current flowing through the transistor and in load resistor $R_L$ is then no longer equal to its original value. This is illustrated mathematically for small EMI signals. Assume that the EMI disturbance is small and sinusoidal, and expressed as follows:

$$v_{emi}(t) = \hat{v} \cdot \sin(\omega t) \tag{4.1}$$

As long as $v_{emi}(t)$ is a small amplitude signal, the voltage $V_s$ at the source of $M_1$ can be written as the sum of a DC term $V_S$ and an AC term $v_s$:

$$V_s = V_S + v_s \tag{4.2}$$

Specifying that $1/g_{m1} \ll R_L$, where $g_{m1}$ represents the transconductance of $M_1$, the transfer function from the EMI source to the source of $M_1$ equals to:

$$H_1(s) = \frac{v_s(s)}{v_{emi}(s)} = \frac{1}{1 + g_{m1} \cdot R_s} \tag{4.3}$$

This expression is valid as long as the amplitude of $v_{emi}(t)$ is small and as long as transistor $M_1$ remains in saturation. Observe that according to the DPI specification, the sum of source resistor $R_s$ and protection resistor $R_p$ (not shown on the schematic) varies between 50 $\Omega$ and 150 $\Omega$. This signifies that under normal circumstances, $R_s \ll 1/g_{m1}$, meaning that $|H_1(s)| \approx 1$. As long as $M_1$ is in saturation and using relation (4.2), output current $I_d$ is expressed as follows [San06]:

$$I_d = \frac{\mu C_{ox}}{2} \cdot \frac{W}{L}(V_G - V_S - v_s(t) - V_t)^2 \tag{4.4}$$

The average over time of drain current $I_d$ can be calculated using the following formula [Gly96]:

$$\overline{I_d} = \lim_{T \to \infty} \frac{1}{T} \int_{\frac{-T}{2}}^{\frac{T}{2}} I_d(t) \cdot d(t) \tag{4.5}$$

In the event that there is no EMI, the average drain current is equal to the DC bias current itself, namely:

$$\overline{I_{d\_noEMI}} = \frac{\mu C_{ox}}{2} \cdot \frac{W}{L}(V_{GS} - V_t)^2 \tag{4.6}$$

[1] Recall that the extra protection resistance $R_p$ which is added in series with $R_s$ may vary between 0 and 100 $\Omega$ [DPI].

In case EMI is present and if the EMI amplitude is small enough so that linear analysis can be used, the eigenfunction property of linear time invariant systems can be applied [Dut97]. Substituting $v_{emi}(t)$ as expressed in (4.1), and using the inverse Laplace operator on (4.3), yields:

$$v_s(t) = \mathcal{L}^{-1}\{H_1(s) \cdot v_{emi}(s)\} = |H_1(j\omega)| \cdot \hat{v} \cdot \sin(\omega t + \phi) \tag{4.7}$$

where $|H_1(j\omega)|$ is the magnitude an $\phi$ the phase of transfer function $H_1(s)$. The average drain current is then found by substituting (4.7) in (4.4):

$$\overline{I_{d\_withEMI}} = \overline{\frac{\mu C_{ox}}{2} \cdot \frac{W}{L}(V_G - V_S - |H_1(j\omega)| \cdot \hat{v} \cdot \sin(\omega t + \phi) - V_t)^2} \tag{4.8}$$

In accordance with (3.22), three terms can be distinguished when (4.8) is expanded, namely a DC term equaling (4.6), a second term which is linearly proportional to $v_{emi}$ (and whose mean value is consequently zero) and a third term expressing a quadratic proportionality between the drain current and $v_{emi}$. This is further simplified to:

$$\overline{I_{d\_withEMI}} = \overline{I_{d\_noEMI}} + \overline{\frac{\mu C_{ox}}{4} \cdot \frac{W}{L}(|H_1(j\omega)| \cdot \hat{v})^2} \tag{4.9}$$

The latter is evidently not equal to (4.6) and strongly differs from it by the EMI induced quadratic term. This term is the so-called DC shift term, namely, the DC voltage is shifted to a different value owing to the presence of EMI [2]. Observe that the generated DC shift which is derived using a small signal analysis is proportional to the squared amplitude of the EMI signal, and to the squared magnitude of transfer function $H_1(j\omega)$ describing the EMI coupling from the EMI source to the source of $M_1$.

For large EMI amplitudes, transistor $M_1$ is periodically forced into cut-off, and as a result, introduces strong nonlinear components in the expression of $I_d$. Equation (4.9) does not represent the DC shift correctly in that case, and different mathematical approaches must be used to calculate the DC shift. As explained in the previous chapter, power and Volterra series may not be used when the transistor is brutally turned off and on, because the series in question does not converge any longer [Wam98]. This means that other mathematical methods must be used in this case, like orthogonal functions and Wiener functionals [Sch80]. Allegedly, closed loop expressions taking the clipping into account are quite complex to obtain and are equally not all that accurate. In any case, since these computations do not lead to a straightforward circuit solution and since they do not contribute to a better understanding of the originating phenomena, they are not developed further in this work. The bottom line is that DC shift is quickly worsening for increasing EMI amplitudes, owing to the clipping of $M_1$ and other strong nonlinear distortion effects.

---

[2] The attention of the reader is drawn to the observation expressed in page 51.

## 2.2 Common-source output circuits

The second class of output circuits couples the output signal to the drain of one or more transistors. These structures are particularly suitable when a high gain is required as well as a high output impedance, e.g. in operational transconductance amplifiers. In addition, drain outputs very often allow a rail-to-rail operation, which is not possible when source followers are used owing to the latter's poor driving capability [Raz01]. Finally, since source followers tend to have a poor noise performance and consume current without providing voltage gain, they tend to be avoided and replaced by common-source transistors in low noise and low power applications. For this reason, many class AB output stages have two output transistors which are connected drain-to-drain, e.g. as in [Wu94]: many variations on this structure have been developed, and are commonly used in contemporary integrated circuit design [Alo05]. Consider as an example the well-known common-source configuration (Fig. 4.2a). This circuit is used in combination with a load (pull-up), which in an integrated environment is often realized as a high impedance current source or as a complementary PMOS transistor. Again, for the purpose of generality, an explicit load resistor $R_L$ is considered here. Depending on how the gate of the output transistor is driven, the EMI signal may couple through the parasitic drain-gate capacitance ($C_{dg}$) to the gate of the transistor. The EMI source is once again represented by an EMI voltage source with an internal resistance $R_s$ which is coupled to the circuit using a coupling capacitor $C_c$, according to the DPI specification [DPI]. This setup equally generates a net DC shift in the drain current, as is calculated here [Red07b]: refer to Fig. 4.2b. Assume transistor $M_1$ is working in saturation, and that the EMI source is sinusoidal, and expressed as in (4.1). As long as $R_L \gg R_s$, the EMI source is fully coupling in

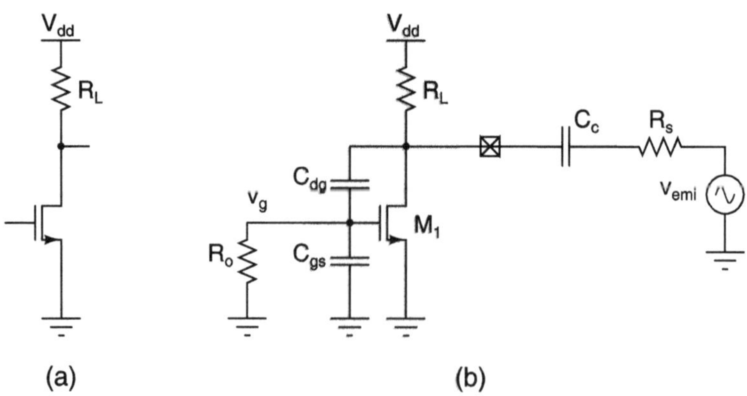


*Figure 4.2.* (a) Common-source amplifier. (b) Common-source amplifier with EMI injection at its output.

on the drain of $M_1$. If the EMI signal amplitude is small enough so a linear analysis can be used, the transfer function from $v_{emi}$ to the gate of $M_1$ ($v_g$) is specified as:

$$H_2(s) = \frac{v_g(s)}{v_{emi}(s)} = \frac{C_{dg} \cdot R_o \cdot s}{1 + s \cdot R_o \cdot (C_{dg} + C_{gs})} \tag{4.10}$$

Parallelly to the previous section, the average value of $I_d$ is now calculated. In case there is no EMI, the average value of the drain current is equal to (4.6). In case EMI is present and if the EMI amplitude is small enough so that linear analysis can be used, the same reasoning as in (4.7) and (4.8) can be followed, yielding an average drain current which is now equal to:

$$\overline{I_{d\_withEMI}} = \overline{I_{d\_noEMI}} + \overline{\frac{\mu C_{ox}}{4} \cdot \frac{W}{L}(|H_2(j\omega)| \cdot \hat{v})^2} \tag{4.11}$$

As expected, the latter is evidently not equal to (4.6) owing to the DC shift term. For higher EMI amplitudes, (4.11) does not express the DC shift correctly any longer and other mathematical approaches must be used, as mentioned earlier. However, the small signal approach is valid much longer for common-source type output circuits compared to common-drain type output circuits, because $M_1$ stays in saturation for much higher EMI swings since it essentially behaves as a current source. This is clarified further on.

## 2.3 Comparing the electromagnetic susceptibility

As can be seen from a comparison between (4.9) and (4.11), small signal calculations yield that the EMI induced DC shift of common-source as well as common-drain output circuits are equal to each other, the only difference residing in the transfer functions $H_1(j\omega)$ and $H_2(j\omega)$. The Bode plot depicting the magnitudes of both transfer functions is represented schematically in Fig. 4.3. Four important observations can be made with respect to this plot:

- $|H_1(j\omega)|$ is independent of the frequency, and its magnitude is (much) higher than the maximal value of $|H_2(j\omega)|$. Indeed, the EMI source resistance $R_s$ is specified equal to 50 Ω in the DPI specification and can be increased up to 150 Ω if a series protection resistor $R_p$ is included [DPI]: however, $g_{m1}$ is typically not larger than a few mS at best, and so this means that $|H_1(j\omega)|$ is quite close to unity under normal circumstances.

- On the other hand, the maximal value of $|H_2(j\omega)|$ is defined by the parasitic capacitive coupling from the drain to the gate of $M_1$. This coupling can be minimized by decreasing $C_{dg}$ as observed in (4.10). In any case, $C_{dg}$ is typically much smaller than $C_{gs}$, since the former purely consists of the overlap capacitance, while the latter includes two thirds of the channel capacitance as well [Raz01].

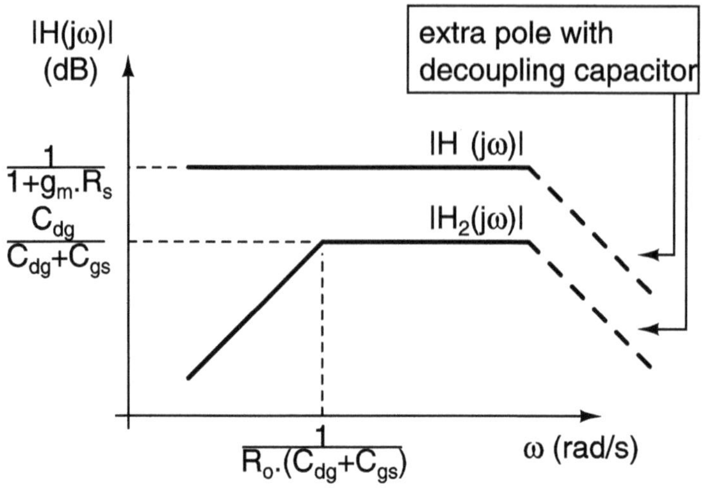


*Figure 4.3.* Bode plots of $|H_1(j\omega)|$ and $|H_2(j\omega)|$.

- Moreover, $H_2(j\omega)$ contains a pole which can be warped to very high frequencies, depending on the impedance driving the gate of $M_1$ ($R_o$). This pole frequency can be increased by reducing $R_o$.

- In both circuits, a decoupling capacitor at the circuit output filters the EMI, and steadily improves the EMI behavior of the respective output stages at higher frequencies. This is shown in the Bode plot depicted in Fig. 4.3. The decoupling capacitor helps to filter out-of-band EMI signals, and should not interfere with the wanted signal spectrum. Since integrating on-chip capacitors is extremely area consuming (and therefore, quite costly), it is important to realize a given pole with a capacitor that is as small as possible. In order to place the pole at an angular frequency equal to $\omega_p$, the decoupling capacitor added to the common-drain circuit must be approximately equal to:

$$C_{d\_CD} = \frac{g_{m1}}{\omega_p} \tag{4.12}$$

While in order to realize the same pole at the output of the common-source circuit, the needed capacitor is given by:

$$C_{d\_CS} = \frac{1}{R_L \cdot \omega_p} \tag{4.13}$$

Since $R_L$ is in practice much higher than $1/g_{m1}$, the decoupling at the output of the common-source circuit requires a smaller capacitor in order to realize the same high frequency pole. This is again not very surprising, considering the output impedances of a common-source and a common-drain output circuit.

*Example*: In this paragraph, the maximal EMI induced DC drain current shifts appearing in a typical common-source and common-drain output circuit are assessed and compared using a standard 0.35 μm CMOS process. First of all, the expressions for the maximal EMI induced DC shift of the drain current for a common-drain ($I_{d\_EMI\_MAX\_shift\_CD}$) and a common-source stage ($I_{d\_EMI\_MAX\_shift\_CS}$) are summarized here:

$$\overline{I_{d\_EMI\_MAX\_shift\_CD}} = \overline{\frac{\mu C_{ox}}{4} \cdot \frac{W}{L} \cdot \left( \frac{\hat{v}}{1 + g_{m1} \cdot R_s} \right)^2} \tag{4.14}$$

$$\overline{I_{d\_EMI\_MAX\_shift\_CS}} = \overline{\frac{\mu C_{ox}}{4} \cdot \frac{W}{L} \cdot \left( \frac{C_{dg} \cdot \hat{v}}{C_{dg} + C_{gs}} \right)^2} \tag{4.15}$$

According to the DPI specification, the EMI source impedance $R_s$ is equal to 50 Ω. Substituting $g_{m1} = \mu C_{ox} \cdot \frac{W}{L}(V_{GS} - V_t)$ in (4.14), yields the following relation:

$$\overline{I_{d\_EMI\_MAX\_shift\_CD}} = \overline{\frac{\mu C_{ox}}{4} \cdot \frac{W}{L} \cdot \left( \frac{\hat{v}}{1 + 50 \cdot \mu C_{ox} \cdot \frac{W}{L}(V_{GS} - V_t)} \right)^2} \tag{4.16}$$

Assuming that $M_1$ is working in saturation, its parasitic drain-gate and gate-source capacitances are approximated by the following expressions [Raz01]:

$$C_{gs} = W \cdot C_{gso} + \frac{2}{3} \cdot W \cdot L \cdot C_{ox} \tag{4.17}$$

$$C_{dg} = W \cdot C_{gdo} \tag{4.18}$$

where $C_{gso}$ and $C_{dgo}$ represent the gate-source and gate-drain overlap capacitances per transistor width and where $C_{ox}$ is the channel capacitance per channel area. If (4.17) and (4.18) are substituted in (4.15), and assuming that the overlap capacitances are equal to each other, this gives:

$$\overline{I_{d\_EMI\_MAX\_shift\_CS}} = \overline{\frac{\mu C_{ox}}{4} \cdot \frac{W}{L} \cdot \left( \frac{C_{dgo} \cdot \hat{v}}{2 \cdot C_{dgo} + \frac{2}{3} \cdot L \cdot C_{ox}} \right)^2} \tag{4.19}$$

Since both expressions (4.16) and (4.19) are as a function of the transistor length, they can be represented and compared in the same plot: refer to Fig. 4.4. In the upper part of this plot, the drain current shift of a common-source and a common-drain stage have been plotted versus the transistor length using the transistor width as a parameter. In the lower part of this same plot, the drain current shift of both output circuits has been plotted versus the transistor length

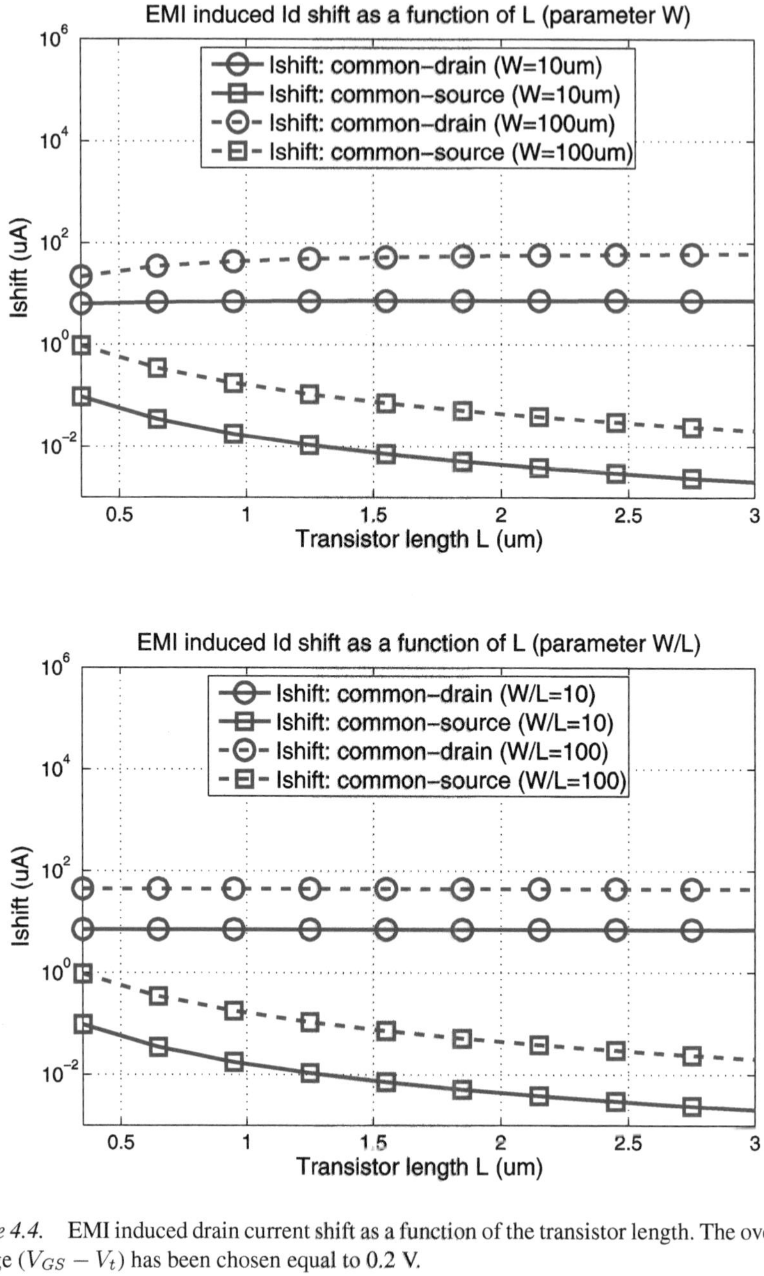


*Figure 4.4.* EMI induced drain current shift as a function of the transistor length. The overdrive voltage ($V_{GS} - V_t$) has been chosen equal to 0.2 V.

using the W/L ratio as a parameter. As can be observed, the EMI induced DC current shift increases for wider transistors, because both (4.16) and (4.19) are directly proportional to the transistor width. Reversely, for an increasing length, the DC current shift of a common-source stage decreases because the gate-source parasitic capacitance is proportional to the transistor length, while the drain-gate capacitance is purely a function of the overlap capacitance times the transistor width, as can be observed in (4.17) and (4.18).

The conclusion is that using basic first order hand calculations, the theoretical DC current shift associated to a common-source stage lies approximately two orders of magnitude beneath the drain current shift associated to a common-drain structure. Moreover, the DC current shift of a common-source stage can further be reduced by increasing the transistor length, because this increases the gate-source capacitance without affecting the drain-gate capacitance. Finally, observe that the curves plotted in Fig. 4.4 are worst case curves, meaning that the DC current shift associated to a common-source stage can be reduced even more by pushing the dominant pole to higher frequencies (e.g. by driving the gate of $M_1$ with a low impedance $R_o$, refer to Fig. 4.3).

## 2.4 Large EMI amplitudes

As stated previously, linear analysis may not be applied for large EMI signals, and expressions (4.9) and (4.11) do not represent the correct DC shift in that case. Indeed, for large EMI amplitudes, the signal transfer from the EMI source to the gate-source voltage of the output circuit may not be expressed using linear transfer functions $H_1(s)$ and $H_2(s)$. However, these transfer functions express the origin of the nonlinearity, and consequently may offer an indication on how the nonlinear behavior evolves for increasing EMI amplitudes. First of all, observe that $H_1(s)$ depends on $g_{m1}$, which is directly impacted by the EMI disturbance since the latter is injected directly into the source of $M_1$, hereby continuously altering its gate-source voltage:

$$g_{m1} = \mu C_{ox} \cdot \frac{W}{L}(V_{gs} - V_t) \tag{4.20}$$

This nonlinear $g_{m1}$ is responsible for a heavy nonlinear drain currents flowing though $M_1$: this nonlinear current generates a major supplementary DC shift, which is not detected using small signal analyses. This phenomenon is in fact similar to what happens in the source follower example described in Chap. 3. Conversely, $H_2(s)$ is solely depending on $C_{gs}$, $C_{dg}$ and $R_o$. The former two are parasitic capacitances, and are of course depending on the biasing and the immediate voltage present at the drain and at the gate of $M_1$. However, they are not subject to change as strongly as $g_{m1}$ in the previous case: furthermore, as explained in the previous chapter, nonlinear capacitors do not contribute to charge pumping. Additionally, $R_o$ is in itself not easily disturbed by EMI since the EMI reaching the gate of $M_1$ is capacitively divided by $C_{dg}$ and $C_{gs}$, and therefore already quite attenuated. So it can be concluded that for large amplitude EMI, a common-source circuit remains linear much longer than a common-drain one. For the same reasons, the corresponding EMI induced DC

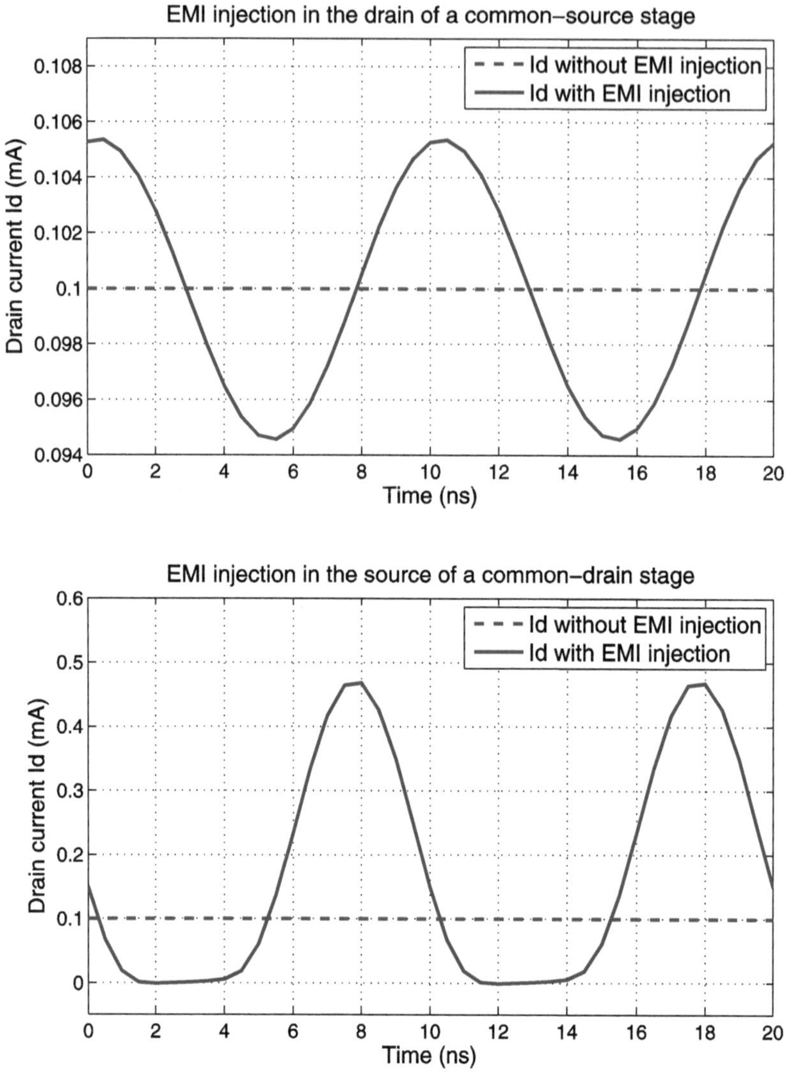


*Figure 4.5.* Drain current in a common-source and common-drain stage when EMI (500 mV at 100 MHz) is injected respectively in the drain and the source of the output transistor.

shift is much smaller for the former than for the latter, if the same EMI source is considered. This is illustrated in Fig. 4.5, where transient simulations of an identical EMI injection in a common-source and a common-drain stage are depicted. The amplitude of the EMI source is 500 mV and the frequency equals 100 MHz. The EMI source has been injected conforming to the DPI injection specification (Figs. 4.1c and 4.2b). Observe that the drain current $I_d$ in the common-drain stage appears heavily distorted compared to the drain current in the common-source stage. These simulation results corroborate the theoretical deductions.

It has been shown that a common-drain type of output exhibits a much higher susceptibility towards EMI than a common-source one. Consequently, common-source output circuit types should be used whenever possible. However, this may not necessarily be achieved so easily in practice. Previous observations are used in two case studies, illustrating the EMI behavior of both output circuit types from a practical point of view.

# 3 Case study 1: EMI resisting DC current regulator

## 3.1 EMI issues in a classic DC current regulator

Consider an IC regulator which is trimmed externally with a resistor, depicted in Fig. 4.6a. This circuit generates a constant DC current ($I_d$) whose value is set by an external trimming resistance ($R_L$) and an internally generated fixed reference voltage ($V_{REF}$). A current mirror copies and scales this current so that it can be used internally to bias subsequent stages of the IC in question ($I_{ref\_1}, I_{ref\_2}, \ldots, I_{ref\_n}$). The main disadvantage of this circuit is that the gate-source voltage of $M_1$ is nonlinearly dependent on its drain current due to the fact that the transconductance of $M_1$ is finite. Because of this prop-

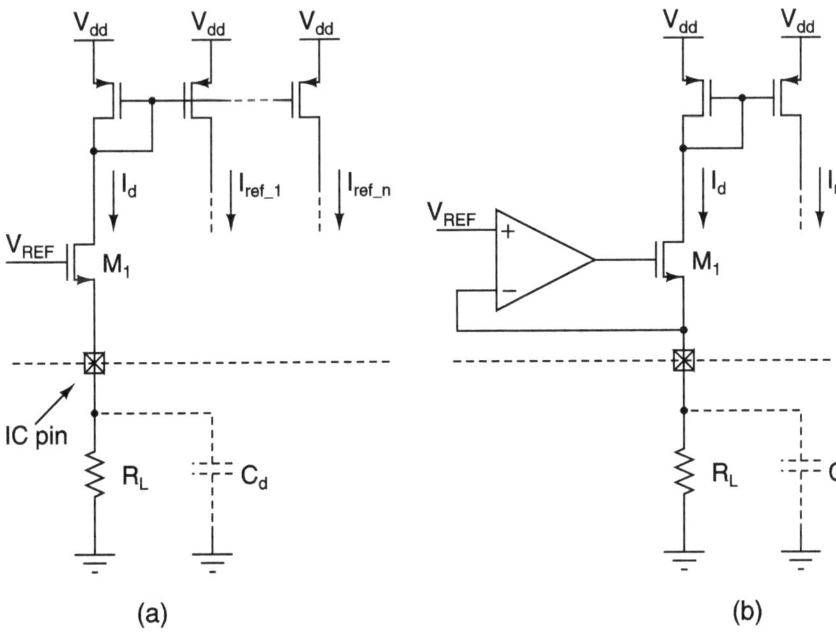


*Figure 4.6.* Basic resistively trimmed DC current regulator (a) Without opamp. (b) With opamp.

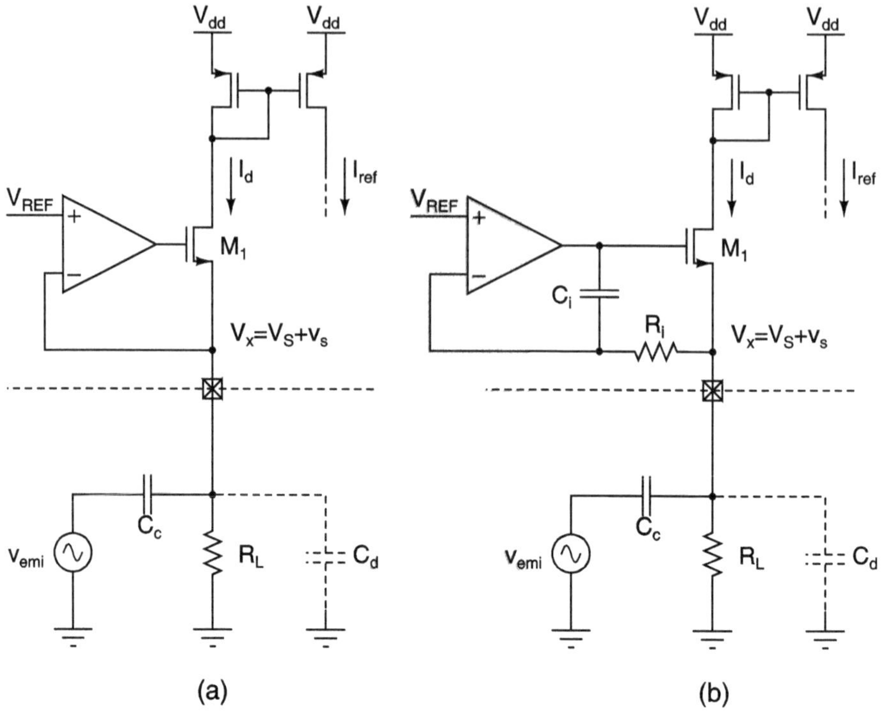


*Figure 4.7.* Current regulator (a) With EMI injection model. (b) With Miller integrator to reduce the GBW of the opamp.

erty, the slightest variation of the transistor parameters (e.g., owing to processing, temperature, etc.) causes the generated current $I_d$ to shift to another value that satisfies the bias conditions. A prevalent solution is to decrease the impedance seen at the source of $M_1$, or in other words, to make $M_1$ a better voltage follower. This can be realized using an opamp, as shown in Fig. 4.6b: the opamp linearizes transistor $M_1$ and increases its transconductance. This circuit is based on the principle of a classic series voltage regulator [Gra01]. Depending on the opamp gain, a very precise and trimmable reference current is obtained this way.

This circuit works fine as long as no EMI disturbance is present. However, it is very susceptible to EMI which is injected into its trim pin, as mentioned in [Cas04]. This predisposition is illustrated as follows: refer to Fig. 4.7a. If EMI is injected on this trim pin (modeled by EMI disturbance source $v_{emi}$ and coupling capacitor $C_c$, according to the DPI specification [DPI]), the drain current $I_d$ will contain undesirable EMI induced components. These propagate to $I_{ref}$, and consequently, to the subsequent stages that are biased by this reference current, thereby possibly disrupting their correct operation. It is, therefore, very important to remove and filter these EMI components before they manage to reach and mix with $I_{ref}$. Furthermore, and this is much more difficult to achieve, these EMI components must be filtered in a highly linear way, so as to avoid as much as possible the detrimental DC shift phenomenon, which was derived and explained in Chap. 3. Since such a current regulator is categorized as a "zone 3" appliance in the DPI specification (refer to Table 2.2 in Chap. 2), the maximum forward power that must be injected into the circuit is equal to 50 mW [DPI]. The internal resistance of the EMI source is equal to 50 Ω, but since this is such a small value compared to the trimming resistance $R_L$, this resistor has been omitted from the schematic in order not to overload it unnecessarily. However, this source resistance increases the filtering at higher EMI frequencies, which in turn improves the circuit operation. In addition, the self-inductances of bondwires and IC leads provide more filtering at higher frequencies, which equally improves the circuit operation. Externally, it is strongly recommended (and depending on the injected EMI level, even mandatory) to decouple the trim pin by means of an external capacitor ($C_d$): however, for the sake of the argument, let's assume at first that, since an EMI issue is present in this circuit, this decoupling capacitor is either absent, either simply ineffective at the respective EMI frequencies. The dimensioning and the effect of $C_d$ will be evaluated further on. Following the approach that was used previously, the small signal operation is derived first, in order to detect the appearing EMI problems. The large signal analysis follows next, focusing on the nonlinear distortion, and particularly on the resulting EMI induced DC shift.

### 3.1.1 EMI issues: small signal analysis

First of all, assume that the EMI disturbance is small and that it is sinusoidal, so that a linear analysis can be used: $v_{emi}$ is consequently expressed as:

$$v_{emi}(t) = \hat{v} \cdot \sin(\omega t) \tag{4.21}$$

Considering that the opamp can be modeled as a perfect one-pole system, its open loop gain is expressed as:

$$A(s) = \frac{A_{DC}}{1 + \frac{s}{p_1}} \tag{4.22}$$

where $A_{DC}$ is the opamp DC gain and $p_1$ the dominant opamp pole. As long as the amplitude of $v_{emi}$ is small, the voltage $V_x$ at the source of $M_1$ can be written as the sum of a DC term $V_S$ and an AC term $v_s$:

$$V_x = V_S + v_s \tag{4.23}$$

The second term in (4.23) is the small signal component that is induced on the IC trim pin by the EMI source. Observe that this is quite similar to the approach

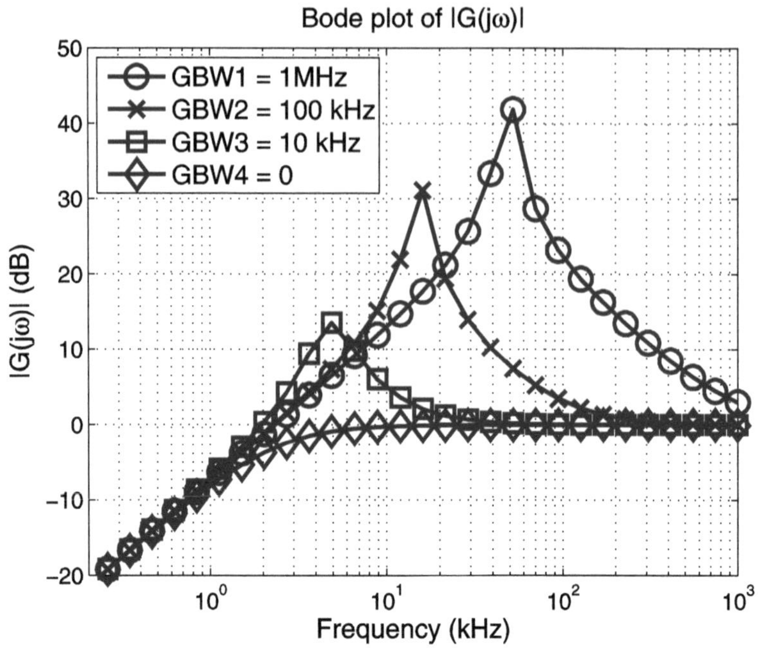


*Figure 4.8.* Bode plot representing the effect of peaking. Notice the higher peaking for increasing GBW. $A_{DC} = 1000$, $g_{m1} = 100$ µS, $C_c = 6.8$ nF.

followed in Sect. 2.1. However, the calculation of the transfer function from $v_{emi}$ to the source of $M_1$ is somewhat complicated by the presence of the negative feedback. Specifying that $1/g_{m1} \ll R_L$ (where $g_{m1}$ represents the transconductance of transistor $M_1$), and provided that $C_d$ is absent, the transfer function representing the coupling between the EMI source and the source of $M_1$ is equal to:

$$G(s) = \frac{v_s(s)}{v_{emi}(s)} = \frac{\frac{C_c}{g_{m1} \cdot p1} \cdot s^2 + \frac{C_c}{g_{m1}} \cdot s}{\frac{C_c}{g_{m1} \cdot p1} \cdot s^2 + \frac{1}{p1} \cdot s + \frac{C_c}{g_{m1}} \cdot s + 1 + A_{DC}} \tag{4.24}$$

This expression is valid as long as the amplitude of $v_{emi}$ is small and as long as transistor $M_1$ remains in saturation. Owing to the complex pole pair in the denominator, peaking takes place for lower EMI frequencies, as shown in Fig. 4.8. This peaking is caused by the EMI disturbance reaching the source of $M_1$ and equally – after being amplified by the opamp – the gate of $M_1$. Needless to say that peaking is very unwanted since it boosts the EMI signal in a certain frequency range. A possible solution to circumvent this effect is to decrease the opamp gain bandwidth product (GBW) by connecting the opamp as a Miller integrator (Fig. 4.7b). Indeed, in the event that the opamp

GBW is smaller than the lowest EMI frequency (150 kHz, according to the DPI specification [DPI]), the gate of $M_1$ can be considered as free of EMI. Mathematically, this means that as long as $GBW = p_1 \cdot A_{DC} \ll \frac{g_{m1}}{C_c}$, transfer function $G(s)$ (4.24) simplifies to:

$$G(s) = \frac{v_s(s)}{v_{emi}(s)} = \frac{\frac{C_c}{g_{m1}} \cdot s}{\frac{C_c}{g_{m1}} \cdot s + 1} \tag{4.25}$$

Which is a transfer function containing one pole and one zero, where there is no peaking. By means of Miller capacitance $C_i$, the dominant pole of the opamp is lowered owing to pole splitting [Gra01]. Adding resistor $R_i$ in the feedback loop forms a classical Miller integrator and completes the circuit, as illustrated in Fig. 4.7b. There is yet another important advantage in decreasing the opamp GBW by means of a Miller integrator, namely, to minimize as much as possible the interference reaching its inverting input. The reason for this will be described in length further on.

### 3.1.2 EMI issues: large signal analysis

Using a Miller integrator to prevent peaking does not solve all the problems, however. Indeed, although the mean voltage at the source of $M_1$ is kept fixed by the DC feedback of the opamp, the drain current of $M_1$ is distorted for increasing EMI amplitudes as shown in Fig. 4.9. The mean value of $I_d$ is therefore not equal to its original value any longer. This is illustrated mathematically for a small EMI amplitude. Assume further that the opamp GBW lies below the lowest EMI frequency owing to the Miller integrator and that the EMI signal appearing at the gate of $M_1$ is negligible. $I_d$ is then expressed as follows:

$$I_d = \frac{\mu C_{ox}}{2} \cdot \frac{W}{L} \cdot (V_G - V_x - V_t)^2 \tag{4.26}$$

where $V_G$ is the DC voltage at the gate of $M_1$, and $V_x$ is represented by the sum of a DC and an AC component as expressed in (4.23). Owing to the Miller integrator, the EMI frequencies are filtered before reaching the gate of $M_1$, and so (4.26) is simplified as:

$$I_d = \frac{\mu C_{ox}}{2} \cdot \frac{W}{L} \cdot (V_{GS} - v_s - V_t)^2 \tag{4.27}$$

Previous expression is similar to the one that was developed in Sect. 2.1, when studying the common-drain output types. The average drain current flowing through $M_1$ is therefore derived from (4.8), and equal to:

$$\overline{I_d} = \frac{\mu C_{ox}}{2} \cdot \frac{W}{L} \cdot (V_{GS} - V_t)^2 + \overline{\frac{\mu C_{ox}}{4} \cdot \frac{W}{L}(|G(j\omega)| \cdot \hat{v})^2} \tag{4.28}$$

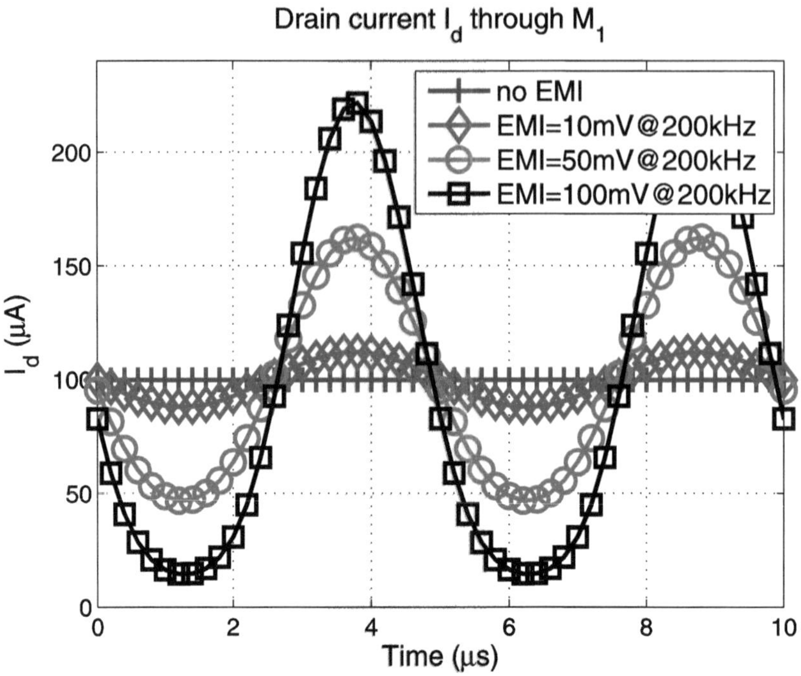


*Figure 4.9.* Distortion effects for increasing EMI amplitudes. GBW of the Miller compensated opamp is 100 kHz.

Previous expression shows the appearing DC shift, since the DC value of the drain current is shifted to a higher value owing to nonlinear effects introduced by EMI. As an example and using the technological parameters of a 0.35 μm process, (4.28) results in an intolerable DC shift in the reference current of more than 12.5%, with an overdrive voltage ($V_{GS} - V_t$) of 0.2 V and an EMI amplitude of scarcely 100 mV. Since the current regulator must withstand an injection of EMI signals in the trim pin of up to 4.4 V in order to comply with the DPI zone 3 measurement specification [DPI], this classic current regulator topology has a very inadequate EMC performance. As stated previously, for even larger EMI amplitudes, (4.28) is not correct any longer, and complex mathematical approaches must be used to calculate the DC shift: however the bottom line is that DC shift quickly worsens for increasing EMI amplitudes. As observed, EMI induced DC shift significantly alters the correct reference current(s) in classic current regulators, depending on the EMI amplitude.

### 3.1.3 Decoupling capacitor $C_d$

Clearly, adding a capacitor which decouples the trim pin helps to attenuate the EMI which is coupled on this node. Depending on the injected EMI levels,

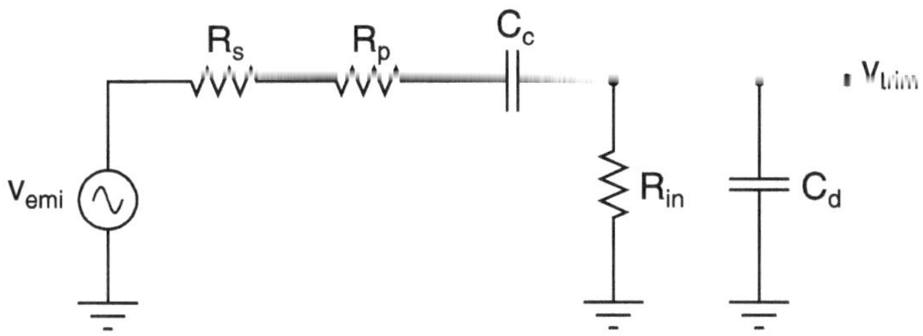


*Figure 4.10.* Simplified EMI model of the classic DC current reference.

and disregarding for a moment the EMI behavior of the current regulator itself, such a capacitor is even mandatory in order to prevent the trim pin from momentarily exceeding the supply rails. In particular, reference voltage $V_{REF}$ is very often generated by a bandgap voltage reference circuit, and lies approximately around 1.2 V. From the moment that the EMI amplitude on the trim pin exceeds $V_{REF}$, the trim node drops below ground during the negative EMI swings, meaning the source of NMOS transistor $M_1$ is pulled below the substrate, hereby forward biasing its source-bulk junction: this may trigger latchup because the parasitic thyristor which is proper to CMOS design is turned on [Has00]. If the EMI amplitude increases even more, the maximum allowable voltages of the IC may be exceeded, and this will break down or destroy the circuit. Along the way, ESD protections connected to this node may (and will) trigger at will, generating a strong nonlinear voltage on this node (refer to Chap. 3). Needless to precise that this situation is highly undesirable. The question is how large this decoupling capacitor must be chosen. Refer to the simplified DPI based model depicted in Fig. 4.10. Clearly, the EMI amplitude which is present on the trim pin depends on:

- The EMI source amplitude, which is maximally equal to 4.4 V for a zone 3 appliance.
- $R_{in}$, representing the parallel combination of $R_L$ and the circuit output impedance seen at the trim pin.
- Decoupling capacitor $C_d$.
- The sum of $R_s$ and $R_p$.

Considering an ideal EMI coupling through $C_c$, the worst case situation is observed at low EMI frequencies (for which $C_d$ presents a high amplitude) and for a zero value of the protection resistor $R_p$. Taking into account a safety margin, the maximal EMI amplitude on the trim pin is defined as 0.75 V. For this value, the necessary $C_d$ is plotted as a function of $R_{in}$ for an EMI frequency

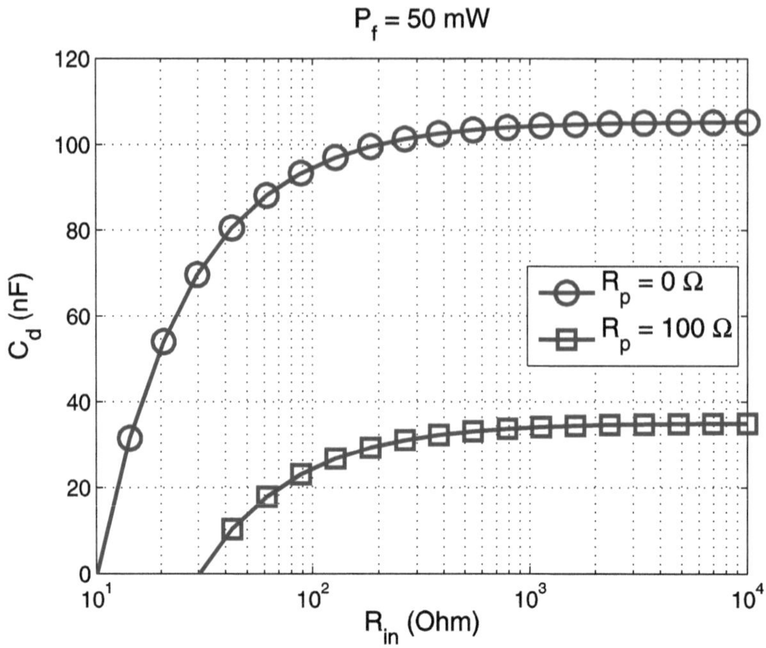


*Figure 4.11.* Decoupling capacitor $C_d$ as a function of $R_{in}$.

of 150 kHz in Fig. 4.11. Observe that if protection resistor $R_p$ is absent, a decoupling capacitor $C_d$ of 100 nF is required. At higher EMI frequencies, the parasitic equivalent series resistor and inductance (ESR, ESL) of $C_d$ prevent this capacitor from decoupling the EMI effectively. This is the reason why a small internal decoupling capacitor must be foreseen as well in order to decouple the trim node at high EMI frequencies.

## 3.2 DC current regulator with a high immunity to EMI

The observations made throughout Sect. 3.1 pinpoint the main weakness of the classic current regulator structure, namely, the fact that the EMI source interferes with the circuit's correct behavior at the source of $M_1$. This is logical, and inherent to the fact that the classic current regulator is a common-drain output circuit. As was derived in Sect. 2, common-drain circuits are much more susceptible to EMI compared to their common-source counterparts. Although the feedback path to the gate of $M_1$ is reasonably inaccessible to EMI thanks to the Miller integrator topology, the source of $M_1$ is directly influenced by the full scale interference which is injected in the trim pin: this results in a DC shift of reference current $I_{ref}$. Even worse, for large EMI voltages, $M_1$ is not biased in saturation any longer, and generates clipping and heavy nonlinear effects.

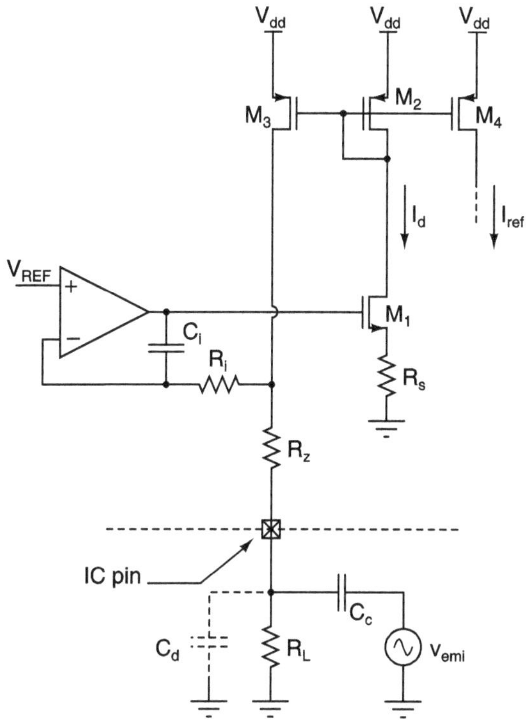


*Figure 4.12.* EMI resisting current regulator.

These problems can therefore be solved by routing the feedback loop in a different way, as shown in Fig. 4.12, in order to realize a common-drain output instead of a common-source one [Red08a]. Transistor $M_1$ is no longer connected directly to the EMI disturbance through its source; instead, it is driven by the opamp, which is connected in a Miller integrator configuration. This Miller integrator creates a low frequency pole, and consequently, the EMI signal reaching the gate of $M_1$ is strongly filtered. A current mirror copies the current and completes the feedback loop while making another (scaled) copy to generate the wanted DC current. If needed, this current can further be filtered by one or more optional EMI filtering current mirror(s) to obtain a smaller ripple on $I_{ref}$ by replacing current mirrors $M_2$-$M_3$ and $M_2$-$M_4$ by EMI resisting ones (refer to Chap. 3). Since the accuracy of the circuit strongly depends on the matching of transistors $M_3$ and $M_4$, these should be designed and layouted accordingly. Owing to the modified feedback loop, the proposed topology introduces a second pole in the closed loop transfer function expression, the low frequency pole being defined by the Miller integrator, and the second pole resulting from decoupling capacitor $C_d$ and trim resistor $R_L$. In order to keep the circuit stable, an internal resistor $R_z$ is added to realize a zero at higher frequencies, hereby cancelling the effect of the second pole. Finally, a small internal decouple capacitor has been connected to the trim pin, in order to overcome the equivalent series resistance (ESR) and equivalent series inductance (ESL) limitations of the external decoupling capacitor $C_d$ as mentioned previously (the latter parasitic elements are not explicitly shown in the schematic). Using this topology, a current regulator achieving a very high EMI suppression is effectively realized. This has major repercussions on the design of the operational amplifier, whose EMC considerations are identified and evaluated here. The effect of EMI interference conveyed to the input terminals of operational amplifiers has been studied and reported in several papers, and is described in detail in Chap. 5. The major effect of EMI in opamps is the presence of a DC offset owing to EMI induced distortion in the input differential pair [Fio01, Red06c]. Focusing on the proposed current regulator design, this means that the interference on the inverting input of the opamp must be preferably as small as possible. Again using linear analysis, the transfer function of the EMI source to the inverting pin of the opamp can be calculated, and split into two transfer functions, namely:

$$H(v_{emi} \rightarrow v_-) = H(v_{emi} \rightarrow v_s) \cdot H(v_s \rightarrow v_-) \tag{4.29}$$

In this expression, $v_s$ is the AC voltage on the trim pin, and $v_-$ is the AC voltage on the inverting input of the opamp. The first transfer function ($H(v_{emi} \rightarrow v_s)$), describing the transfer from the EMI source to the trim pin of the DC regulator circuit, is defined by the amount of EMI coupling (coupling capacitor $C_c$), the presence of a decoupling capacitor ($C_d$), the internal resistance of the EMI source ($R_s$) and the output impedance of the circuit at the trim pin. However, since this coupling occurs outside the IC in question, and since the goal of this current regulator circuit is to provide as much EMI rejection as possible within the IC itself, it will be considered that the disturbance generated by the EMI source appears in full on this trim pin (forming a worst case scenario). The second transfer function ($H(v_s \rightarrow v_-)$), represents the transfer from this trim pin to the inverting input of the opamp. As discussed earlier, this coupling must be minimized at all cost. Assume that the opamp behaves as an ideal one-pole system, as described in (4.22), then the transfer function from the trim pin to inverting opamp input is equal to:

$$H(v_s \rightarrow v_-) = \frac{\frac{s}{p_1} + 1}{\frac{R_i \cdot C_i}{p_1} \cdot s^2 + (R_i \cdot C_i \cdot (A_{DC} + 1) + \frac{1}{p_1}) \cdot s + 1} \tag{4.30}$$

Disregarding the high frequency pole, (4.30) simplifies to:

$$H(v_s \rightarrow v_-) \approx \frac{\frac{s}{p_1} + 1}{R_i \cdot C_i \cdot (A_{DC} + 1) \cdot s + 1} \tag{4.31}$$

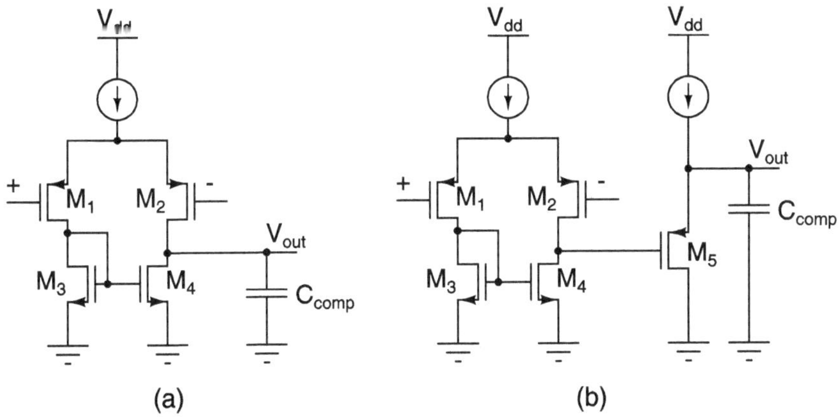


*Figure 4.13.* Single stage opamp (a) With actively biased differential pair. (b) Buffered with a source follower to decrease the output resistance.

Equation (4.31) shows that this transfer function is composed of the expected Miller pole, and of a zero that is present on the same frequency as the original pole of the opamp. In order to reduce the EMI signal on the inverting input of the opamp, the integrator pole must, therefore, be moved to lower frequencies, while the zero must be shifted toward higher frequencies.

Suppose that the opamp is composed of a single differential pair that is actively biased, as shown in Fig. 4.13a. Assuming that its internal parasitic poles are small, then its dominant pole $p_1$ can be expressed as $p_1 = 1/(r_o \cdot C_{comp})$, where $r_o$ is the equivalent resistance of the parallel combination of the output resistances of $M_2$ and $M_4$ and where $C_{comp}$ is the compensation capacitor [Lak94]. In present case, $C_{comp}$ is equal to $C_i$ owing to the Miller theorem, provided that the opamp open-loop gain is large [Raz01]. Equation (4.31) can then be rewritten as:

$$H(v_s \rightarrow v_-) = \frac{r_o \cdot C_i \cdot s + 1}{R_i \cdot C_i \cdot (A_{DC} + 1) \cdot s + 1} \tag{4.32}$$

The pole and zero frequencies of (4.32) are equal to:

$$\begin{cases} \omega_p = \frac{1}{R_i \cdot C_i \cdot (A_{DC}+1)} \\ \omega_z = \frac{1}{r_o \cdot C_i} \end{cases} \tag{4.33}$$

So, in order to decrease the integrator pole without moving at the same time the location of the high frequency zero, $R_i$ must be increased and not $C_i$, since increasing $C_i$ causes the pole as well as the zero to move to lower frequencies, while increasing $R_i$ decreases the integrator pole without affecting the high frequency zero. Another approach is to use a source follower to substantially lower the output impedance of the opamp, as depicted in Fig. 4.13b. This way,

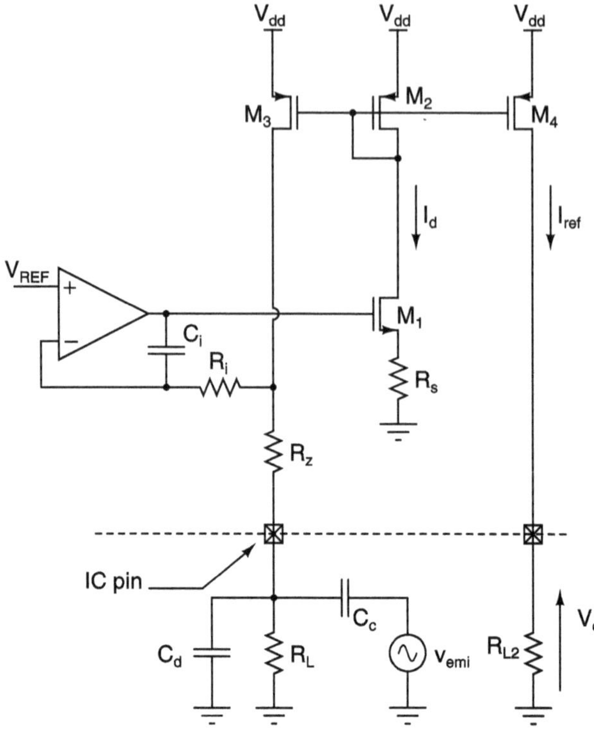


*Figure 4.14.* EMI resisting current regulator as implemented in the test-IC.

the zero is moved to higher frequencies (in this case: $\omega_z = g_{m5}/C_i$, where $g_{m5}$ represents the transconductance of $M_5$).

## 3.3 Measurements

The proposed current regulator (depicted in Fig. 4.14) has been integrated using the AMIS 0.35 μm CMOS technology [3]. Observe in present schematic that the output node (the drain of $M_4$) has been connected to an external sense resistor $R_{L2}$: this way, $I_{ref}$ is consistently derived by measuring the voltage across $R_{L2}$. A microphotograph of the designed current regulator is shown in Fig. 4.15. The size of the test-IC is 1.1 mm × 1.2 mm. The circuit has been designed to generate a reference current $I_{ref}$ of 100 μA using a reference voltage $V_{REF}$ equal to 1 V. The absolute accuracy of $I_{ref}$ has been specified to 3% and this for EMI frequencies ranging from 300 kHz onwards. Both external resistors ($R_L$ and $R_{L2}$) were chosen equal to 10 kΩ with a tolerance of 1%. The circuit has been measured according to the DPI specification (refer

[3] Processed in AMIS I3T80, using the low-voltage components [Moe02].

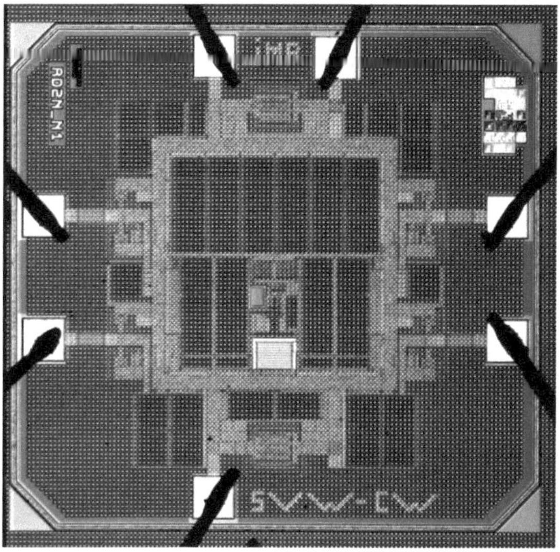

*Figure 4.15.* Microphotograph of the EMI resisting current regulator. The size of the circuit is approximately 1.1 mm × 1.2 mm.

to Chap. 2), as illustrated in Fig. 4.16 [DPI]. The forward power of the EMI source equals +16 dBm. The design values are summarized in Table 4.1. Note that the worst case situation arises at low EMI frequencies, because the Miller integrator is not effectively filtering out the EMI disturbances yet. The corresponding Miller pole lies around 30 Hz and the opamp DC open loop voltage gain is 66 dB. This is illustrated in Fig. 4.17, where EMI is injected in the trim pin at 300 kHz. Although there is a small ripple, the absolute value is still within the 3% accuracy specification. For higher frequencies, this ripple is completely filtered, as can be observed in Fig. 4.18. More importantly, the mean value stays exactly the same for all EMI frequencies. The complete set of measurements has been summarized in Table 4.2. These measurements confirm that there is no DC shift, and this up to the highest measured frequency of 3 GHz. This is a huge improvement compared to the 12.5% DC shift obtained with an EMI amplitude of 100 mV using the classic current regulator, as was theoretically derived on page 100.

## 4 Case study 2: EMI resisting LIN driver

The demand for reliable automobile electronics technology is soaring, since many vehicles incorporate large amounts of electronic control systems [Ale08, Arm08, Wie06]. Because all these systems have to communicate with each other, a lot of wiring is necessary to connect and interconnect these various applications, with the penalties of an increased cost as well as a limited diagnos-

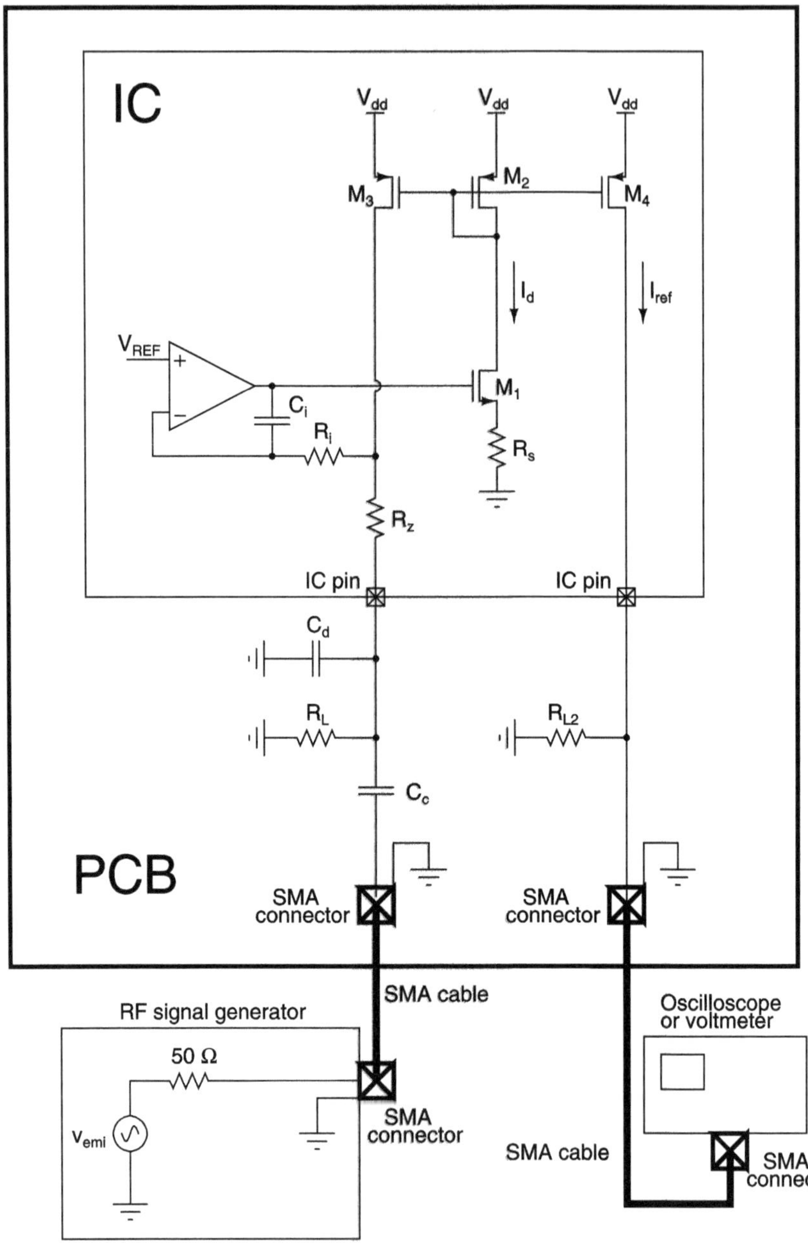


*Figure 4.16.* Direct power injection measurement setup, as described in the DPI specification [DPI].

*Table 4.1.* Design values and specifications.

| | |
|---|---|
| $V_{REF}$ | 1 V |
| $I_{ref}$ | 100 μA |
| $I_d$ | 100 μA |
| $R_L = R_{L2}$ | 10 kΩ |
| Max. injected EMI level | +16 dBm |
| $I_{ref}$ absolute accuracy | 3% |
| $C_d$ | 100 nF |
| $C_d$ ESR | 50 mΩ |
| $C_d$ ESL | 63 nH |

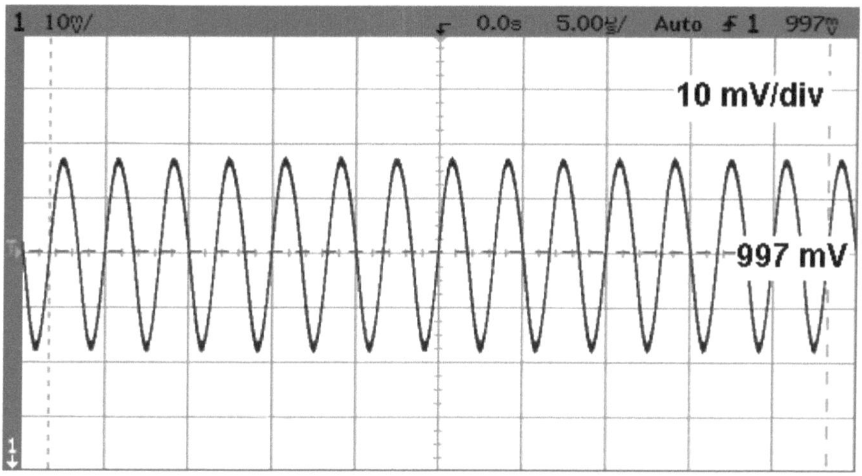


*Figure 4.17.* Worst case measured output voltage measured across $R_{L2}$, when an EMI of +16 dBm is injected at 300 kHz. Observe the small peak to peak ripple of 35 mV, resulting in a total accuracy of 1.75%. More importantly, the mean value of $I_{ref}$ is equal to the value obtained without EMI injection, indicating that there is no DC shift.

tics capability. For these reasons, the German car component provider Robert Bosch Co. introduced the CAN (Controller Area Network) bus, which is a network aiming to replace the complex and expensive traditional wiring [Far99]. The CAN bus is often used to run real-time critical functions as engine management, antilock brakes, cruise control and so on. Because of its high bandwidth requirement, CAN is quite expensive, and it is quite common to replace parts of the CAN bus with smaller and cheaper LIN (Local Interconnect Network) subnetworks, in order to integrate and control sensors and actuators which do not require the high bandwidth offered by CAN in today's cars [Wan05].

LIN is defined as a single-wire, serial communications protocol which is low cost, low speed (maximum transmission speed = 20 kbit/s) and primarily

*Table 4.2.* Measurement results.

| *EMI frequency (MHz)* | *EMI forward injected power (dBm)* | *EMI amplitude (V)* | *Mean value* $I_{ref}$ (μA) |
|---|---|---|---|
| 0 | – | – | 99.80 |
| 0.15 | 16 | 4 | 99.80 |
| 0.3 | 16 | 4 | 99.80 |
| 1 | 16 | 4 | 99.79 |
| 10 | 16 | 4 | 99.79 |
| 100 | 16 | 4 | 99.78 |
| 500 | 16 | 4 | 99.79 |
| 1000 | 16 | 4 | 99.79 |
| 2000 | 16 | 4 | 99.79 |
| 3000 | 16 | 4 | 99.78 |

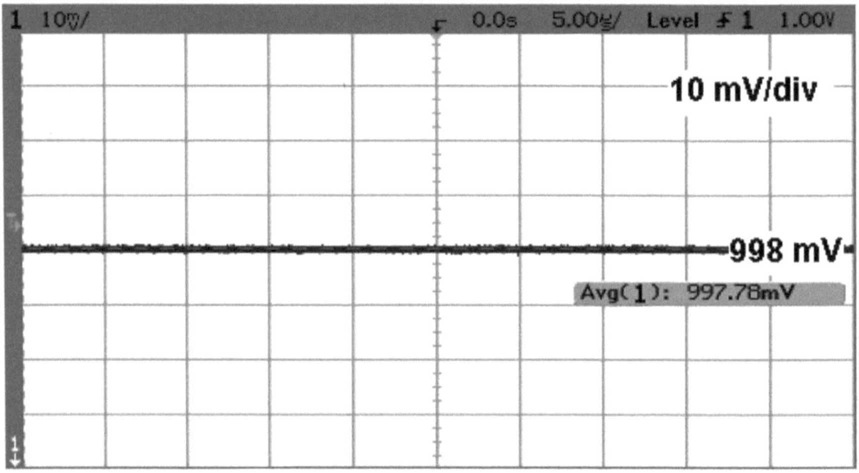


*Figure 4.18.* Output voltage measured across $R_{L2}$, when an EMI of +16 dBm is injected at 100 MHz.

intended to be used for distributed electronic systems in automotive applications [LIN]. Typical applications controlled by LIN include the car door control (windows, door locks, mirrors, . . . ), seats, lighting, rain sensors, intelligent lighting systems and so on [Wan05, Cer05]. Used intertwinedly where necessary, the combined CAN and LIN hybrid networks reduce the global cost to a great extent without affecting the performance of the whole system [Wan05]. Owing to the very stringent EMC requirements concerning automotive electronics [Mey03], the main concept behind the LIN definitions of the physical layer is that it should be able to withstand strong levels of electromagnetic noise, without in turn generating excessive electromagnetic emission which could disturb neighboring circuits. All this makes LIN a valuable communication system, not only in automotive applications, but eventually also in home appliances [Zoe02].

A LIN network is designed to comprise one master node and up to 15 slave nodes (the total number of nodes should not exceed 16 for correct operation). In order to comply with the severe electromagnetic emission standards, requirements have been set on the slope of the output signal of a LIN driver: indeed, the steeper the slope is, the more HF components are generated in the signal spectrum, as has been illustrated in Chap. 2 [LIN]. The LIN specification therefore states that *the signal shape should be carefully selected in order to reduce emissions on one hand and allow for bit rates up to 20 kbit/sec on the other.* For this reason, the slope of the LIN driver output signal must be controlled and set to a convenient value, which emission measurements have shown to lie around 5 μs. This operation is commonly called slew rate control or slope control [Deu03b]. Controlled slew rate output drivers are intensively used in digital output buffers in order to reduce their simultaneous switching noise (SSN, also identified as delta-I or $di/dt$ noise), occurring when the output buffers are switched on at the same time, as well as caused by ground and power supply bounce [Deu03b, Sen93]. Moreover, this slope time must be independent of the battery voltage (which according to the LIN specification may vary between 8 V and 18 V) and of the load (as stated previously, the total number of nodes may vary, and so the correct load is not a priori known) [LIN]. Finally, in order to allow a correct data transmission, the duty cycle of the signal on the LIN bus must not be corrupted by EMI: a summary of allowed deviations of the duty cycle and between which signal levels this value should be measured is included in the LIN specification [LIN]. Moreover, the LIN driver is defined to have an open drain output, with an internal pull-up resistor of 30 kΩ: an external pull-up resistor of 1 kΩ is further added in the master node. A typical LIN transceiver node is depicted in Fig. 4.19.

## 4.1 Classic LIN driver

A classic LIN driver design approach, which provides an output signal with a constant slope time independently of the load and the battery voltage, is depicted in Fig. 4.20a [App05]. This structure is based on the slew rate controlled output buffer technique using capacitive feedback which is described in [Gar98]. Two current sources are intermediately switched on and off, hereby conducting a constant DC current $I_B$ through capacitor $C_f$. A constant current flowing in a capacitor causes a linearly increasing or decreasing voltage across

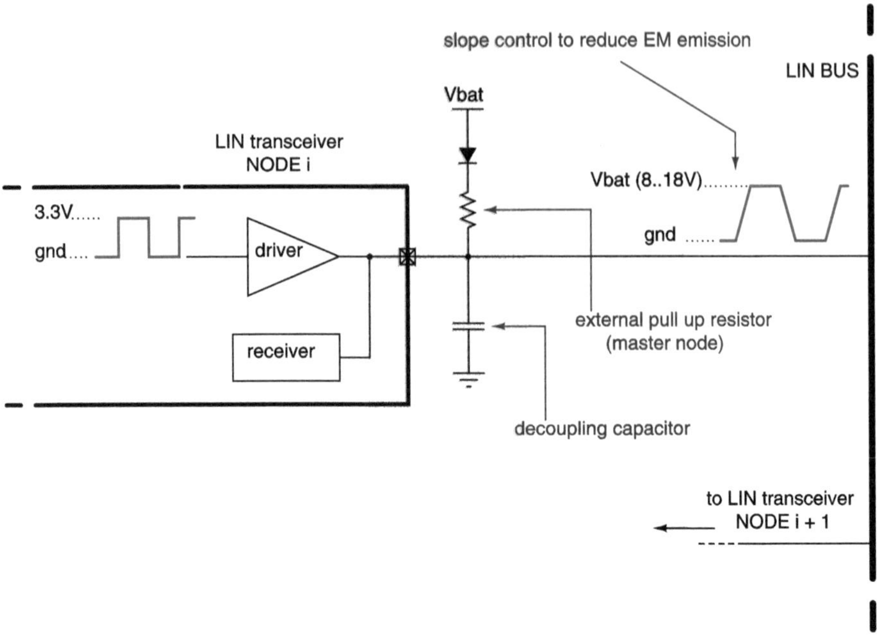


*Figure 4.19.* Overview of a LIN transceiver node.

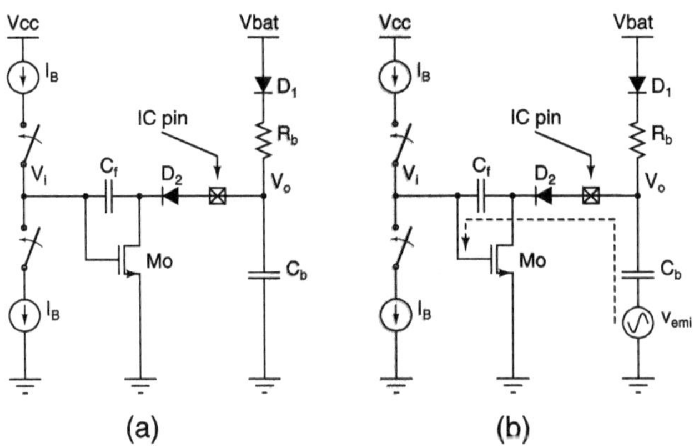


*Figure 4.20.* (a) Classic LIN driver circuit with alternatively switching current sources. (b) Classic LIN driver circuit with EMI injected into the through bus capacitance $C_b$.

the capacitor $(V_{cap}(t))$, depending on the polarity of the current:

$$C_f \frac{d(V_{cap}(t))}{dt} = I_B \rightarrow V_{cap}(t) = \frac{I_B}{C_f}(t - t_0) \qquad (4.34)$$

The voltage gain ($A_v$) of this common-source stage is a function of the total pull-up resistor ($R_b$) and the total bus capacitance ($C_b$), both of which are not precisely known (as explained in the introduction, the exact load connected to the LIN bus may vary). However, as long as $A_v$ is high enough, the output voltage $V_o(t)$ may be expressed as follows:

$$V_o(t) = \left(\frac{|A_v|}{1+|A_v|}\right) \cdot V_{cap}(t) \approx V_{cap}(t) \tag{4.35}$$

Which means that the output voltage is equal to the voltage across the capacitor, as long as the voltage gain is high enough. Additionally, if the current sources delivering $I_B$ are designed to vary proportionally according to the battery voltage, a constant slope time which is independent of the battery voltage is obtained.

This classic LIN driver circuit depicted in Fig. 4.20a performs fine as long as it is not influenced by EMI. Consider Fig. 4.20b. If an EMI disturbance is present on the LIN bus (modeled by voltage source $v_{emi}$ coupling into the LIN bus through bus capacitance $C_b$), it may reach the gate of $Mo$ through $C_f$, hereby possibly corrupting the duty cycle of the signal on the LIN bus. This phenomenon has been reported in [Deu03b], and does not come as a big surprise: indeed, although this output structure is a common-source circuit type (being inherently quite EMI resisting, as explained previously), $C_f$ deteriorates its EMI behavior because the disturbances couple from the LIN bus to the gate of the output transistor [Gie06]. This effect is demonstrated in Fig. 4.21, where it is shown that increasing the EMI amplitude strongly affect the duty cycle of the output signal ($V_o$). Unfortunately, this makes this circuit totally useless in its present form the moment EMI is injected into the LIN bus.

First of all, it is important to note that the LIN driver has a strong semi-nonlinear mode of operation: it is undoubtedly nonlinear because it switches from a high state ($Mo$ off) to a low state ($Mo$ on) and vice versa. However, during the switching and owing to the slope control, output transistor $Mo$ is momentarily biased in its active region, where for small EMI levels a linear operation may be assumed. This previous statement spawns the powerful consequence of allowing a small signal analysis to be performed in this particular condition, in order to gain an intuitive insight in the operation of the circuit. Following this, the same reasoning will be expanded to larger EMI levels and finally to the nonlinear operation mode, where the distinction will be made between a high and a low output level. The advantage of this intuitive technique is that it offers a high transparency in the circuit's operation, despite a strongly reduced mathematical complexity. This approach is viable as long as exact signal levels are not required to be calculated.

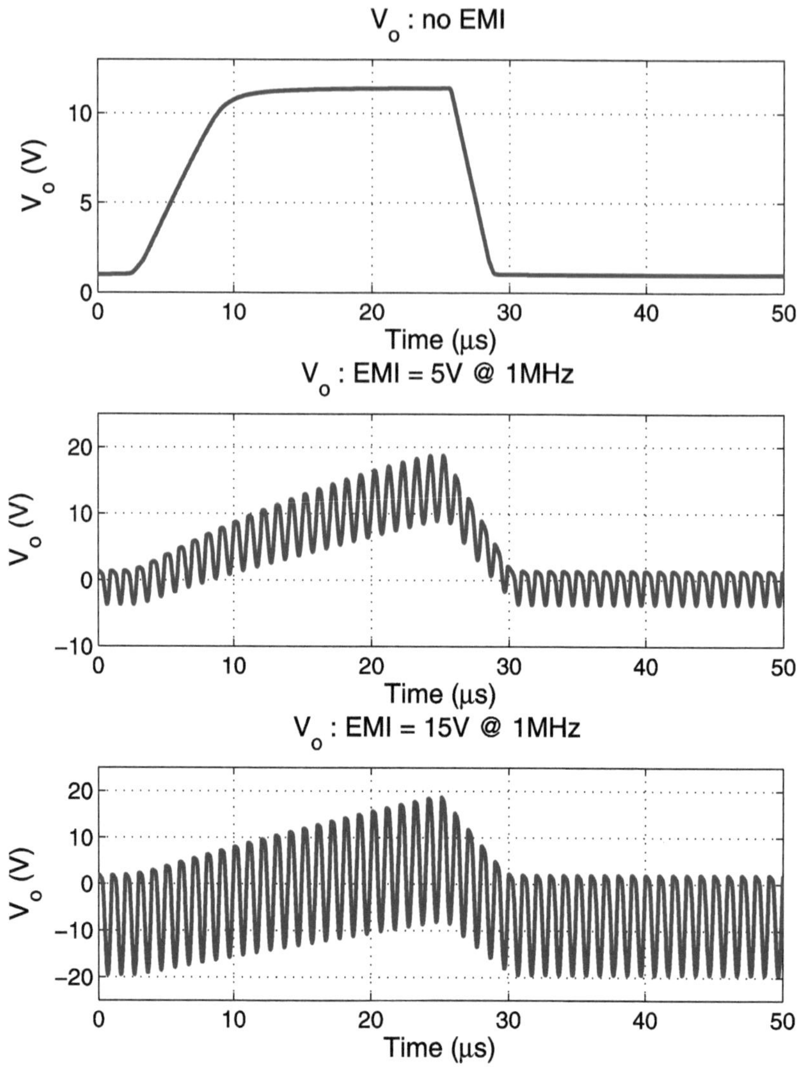


*Figure 4.21.* Distortion of the output signal due to injected EMI: it is not possible to retrieve the duty cycle from the corrupted signal. The stronger the EMI injection, the smaller the mean output voltage during the slope transitions and the high output states.

### 4.1.1 Linear operation mode

- **EMI amplitude is small:** Assume at first that the EMI amplitude level is small. The schematic depicted in Fig. 4.20b is analyzed using small signal analysis. To this end, the EMI source is modeled as a voltage source which is directly injected into the drain of $Mo$ through a source impedance $R_s$

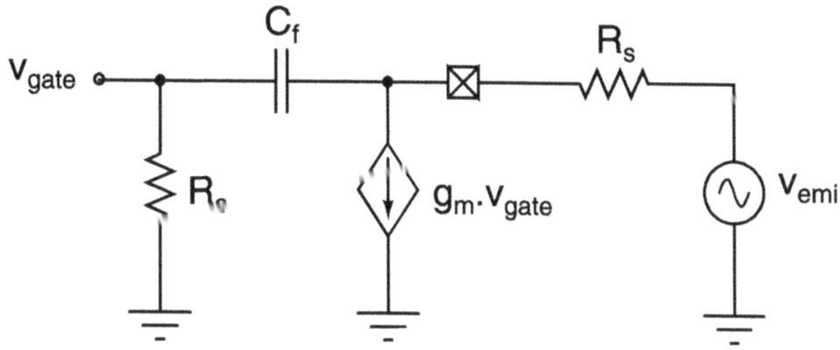


*Figure 4.22.* Classic LIN driver: small signal analysis.

(Fig. 4.22). The following transfer function is obtained:

$$H(s) = \frac{v_{gate}(s)}{v_{emi}(s)} = \frac{s}{s(R_s \cdot g_m + \frac{R_s}{R_o} + 1) + \frac{1}{C_f \cdot R_o}} \tag{4.36}$$

Observe in the previous expression that the parasitic coupling from the drain to the gate of $Mo$ can be reduced on one hand by decreasing $C_f$, which moves the pole to higher frequencies, and on the other hand by decreasing $R_o$, which reduces the magnitude of the maximal achieved coupling. This is exactly the same conclusion as has been drawn while studying the EMI behavior of common-source stages, at the beginning of this chapter. Not surprisingly, the basic LIN driver circuit depicted in Fig. 4.20a is lacking both requirements: it has a relatively high $R_o$, owing to the high output impedance of the current sources and it uses an extra $C_f$, which adds up with the parasitic drain-gate capacitance and promotes the EMI coupling to the gate of $Mo$. It is consequently not surprising that EMI problems tend to appear when this structure is used.

- **EMI amplitude is large:** The arising problems are now examined from a large signal point of view. First consider the average drain current when there is no EMI present. Depending on the gate voltage, a certain average DC drain current flows through $Mo$ which is expressed as ($Mo$ in saturation):

$$\overline{I_{do}} = \lim_{T\to\infty} \int_{-\frac{T}{2}}^{\frac{T}{2}} I_{do}(t)dt = \frac{\mu C_{ox}}{2}\frac{W}{L}(V_{GSo} - V_t)^2 \tag{4.37}$$

The moment a sinusoidal interference reaches the gate of $Mo$ ($v_{gate} = \hat{v}_{gate} \cdot \sin(\omega t)$), this average current becomes equal to:

$$\overline{I_{do}} = \overline{\frac{\mu C_{ox}}{2}\frac{W}{L}(V_{GSo} + v_{gate} - V_t)^2} \tag{4.38}$$

Expanding $I_{do}$ in the same fashion as in Sect. 2.2, the following mean drain current is obtained.

$$\overline{I_{do}} = \frac{\mu C_{ox}}{2}\frac{W}{L}(V_{GSo} - V_t)^2 + \overline{\frac{\mu C_{ox}}{4} \cdot \frac{W}{L} \cdot (\hat{v}_{gate})^2} \qquad (4.39)$$

Which (depending on the disturbance magnitude) yields a higher current compared to the average drain current which is obtained without EMI. This is caused by the second right hand term in (4.39), which is the EMI induced DC shift. This DC shift increases the average output current, and consequently decreases the mean output voltage on the LIN bus: as a result, this is a serious issue when the output voltage is supposed to reach a high state. This effect is clearly (not) appreciated in Fig. 4.21: observe that the stronger the EMI injection, the lower the mean high state voltage of the output signal becomes during the slope transitions.

### 4.1.2 Nonlinear operation mode

- ***Mo* is ON:** Let's now consider what happens when $Mo$ is working in its triode region, hereby shorting the LIN bus to the ground with its on resistance $(R_{DS})$. Due to its high $W/L$ (aimed at sinking a maximum continuous drain source current of 40 mA, this transistor is consequently quite large), $R_{DS}$ is very small and therefore, the disturbance reaching the drain of the output transistor is heavily attenuated. This is illustrated in Fig. 4.23, which models the LIN driver output stage: in the event that the transistor is on, the switch is closed, and the EMI is shorted to ground. Additionally, at low EMI frequencies, diode $D_2$ which prevents the ground node from discharging into the LIN bus in the event the latter drops below the ground potential (e.g. due to EMI or to an uncontrolled loss of battery), ensures that the disturbances reaching the drain of $Mo$ are positive: this means that in any case, should EMI still manage to couple through $C_f$, it will be

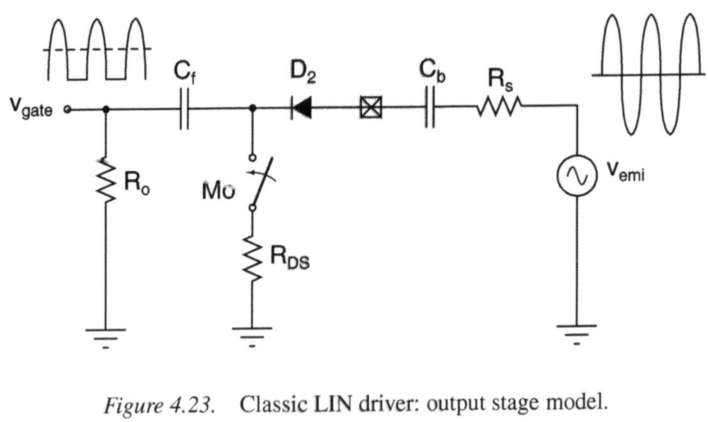


*Figure 4.23.* Classic LIN driver: output stage model.

larger than zero, and the more positive the gate of $Mo$ becomes, the lower the $R_{DS}$ of the output transistor will be, attenuating the output disturbance at the drain even further. Finally, an internal protection must be foreseen so that the drain of output transistor $Mo$ does not fall momentarily below zero during the injection of high frequency EMI, for which diode $D_2$ behaves as a capacitor. Indeed, this would turn on the drain-bulk junction of $Mo$, hereby generating a substrate current flow, with the risk of latching up the device. Luckily, high frequency disturbances coupling into the drain of $Mo$ are considerably attenuated by the external and internal decoupling capacitors (not drawn on Fig. 4.23), as well as by the low on resistance of $Mo$. Consequently, EMI problems are not expected to happen when the output is in a low state.

- ***Mo* is OFF:** This becomes an entirely different matter when the transistor is off: the EMI feedthrough to the drain (and consequently to the gate) of $Mo$ is much higher in this event, because the small $R_{DS}$ which heavily attenuates the disturbance is not present any longer, as seen in Fig. 4.23. Moreover, owing to diode $D_2$, these coupled disturbances are positive, which in turn pumps the average value of the voltage at the gate of $Mo$ to a higher value, hereby increasing the average drain current and consequently decreasing the average voltage level on the LIN bus. This effect is clearly seen in Fig. 4.21, where the high level of the output signal is much more difficult to reach for a higher EMI amplitude.

## 4.2 EMI resisting LIN driver topology: LIN driver 1

The proposed solution circumvents the contradictory need for $C_f$ by splitting the slope control stage from the output, while correcting and adjusting load and battery voltage variations by means of negative feedback using an operational amplifier (Fig. 4.24). The ensuing EMI resisting LIN driver topology has been introduced in [Red06b] and exposed in detail in [Red07a]. The main advantage of this structure is that the extra capacitor connected between the drain and the gate of $Mo$ is left out, and consequently, the coupling between the LIN bus and the sensitive gate of $Mo$ is reduced to its bare minimum. Unfortunately, EMI on the LIN bus couples into the circuit and reaches the gate of $Mo$, namely through the non-inverting opamp input (*path 1*). In addition, the parasitic drain-gate capacitance of $Mo$ $(C_{dg})$ may have a detrimental effect on the operation of the LIN driver if no extra precautions are taken (*path 2*). Both paths are identified in Fig. 4.24, and are closely examined in the following sections to see how their effect can be thwarted using circuit techniques.

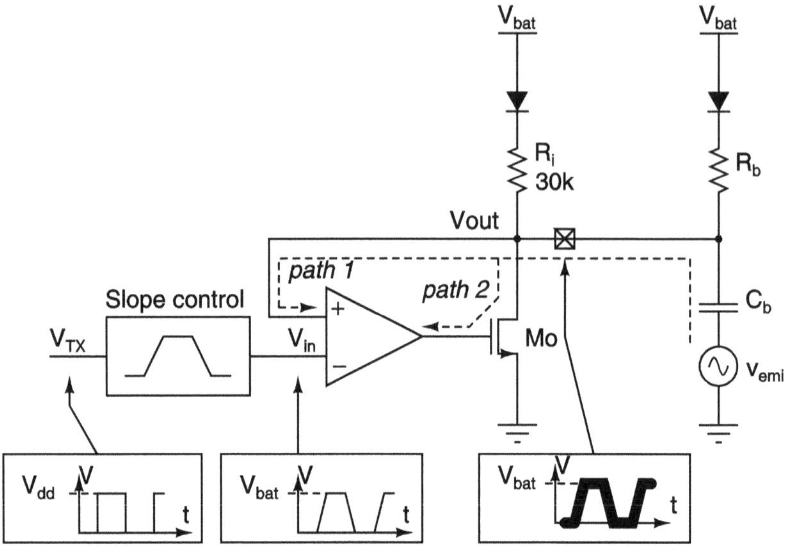


*Figure 4.24.* Basic schematic of the EMI insensitive LIN driver.

### 4.2.1 EMI path 1

- **Introduction:** The non-inverting opamp input is directly connected to the LIN driver output, which means that any disturbance which is present on the LIN bus is amplified by the opamp before appearing at the gate of $Mo$. At first sight, this looks quite unfortunate, to say the least. Moreover, it is not possible to place a filter between the LIN driver output and the non-inverting opamp input, since this filter should filter low EMI frequencies (lowest EMI frequency starts at 150 kHz according to [DPI]), while the dominant pole of this circuit is set by the bus time constant which – according to the LIN specification [LIN] – varies between 1 μs and 5 μs (this corresponds to a pole frequency of respectively 160 kHz and 32 kHz). Clearly, this cancels out the possibility to use linear filtering: however, it leaves the door open for nonlinear filtering as long as the latter does not affect the steady-state behavior of the circuit. This statement will be clarified in the subsequent paragraphs. But first, it will be illustrated that the frequency spectrum of the wanted output signal may be limited to the current dominant pole frequency (varying between 32 kHz and 160 kHz) without any significant loss of information.

- **Frequency spectrum of the output signal:** The wanted output signal on the LIN bus varies between ground and the battery voltage, with a slope time of 5 μs and a duty cycle of 50%. The minimum period of this signal equals 50 μs, since the maximal switching frequency is 20 kbit/s. The spectrum of this signal can be calculated with great accuracy: however, a much quicker calculation method is given in [Goe01], where the frequency spectrum of a trapezoidal wave is normalized, and calculated asymptotically using the pulse width time and the slope time. An illustration of this quick calculation method has already been plotted in Chap. 2 (Fig. 2.6). Using previous worst case values, the frequency spectral component at 32 kHz (the smallest dominant pole frequency) is quickly calculated: it is about 2 times smaller than the normalized DC value, while the frequency spectral component at 160 kHz is 25 times smaller. This means that the wanted output signal on the LIN bus can essentially be considered as a low frequency signal with respect to the LIN driver bandwidth. The consequence of this statement has a major impact on the circuit, as the next paragraph demonstrates.

- **Steady-state error:** For signal frequencies which are inferior to the dominant pole frequency, the signal open loop gain $(T_{signal})$ may be roughly approximated by its DC value $(T_{signal\_DC})$. In its turn, the signal DC open loop gain is equal to the opamp DC gain $(A_{opamp\_DC})$ times the DC voltage gain of $Mo$ $(A_{Mo\_DC})$:

$$T_{signal}(f < 160 \text{ kHz}) \approx T_{signal\_DC} = A_{opamp\_DC} \cdot A_{Mo\_DC} \tag{4.40}$$

Since the wanted output signal is mainly low frequent compared to the LIN driver bandwidth, its instant value is approximately equal to the input signal minus the steady-state error $(V_{SSE})$. The steady-state error of the output signal is proportional to the signal DC open loop gain $(T_{signal\_DC})$. Accounting for a maximal steady-state error of 0.5 V and using the extreme battery voltage values, the following values for the signal open loop gain are obtained:

$$\begin{cases} \epsilon_{min} = \frac{0.5 \text{ V}}{18 \text{ V}} \approx 2.8\% \rightarrow |T_{signal\_DC}| = 31 \text{ dB} \\ \epsilon_{max} = \frac{0.5 \text{ V}}{8 \text{ V}} \approx 6.3\% \rightarrow |T_{signal\_DC}| = 24 \text{ dB} \end{cases} \tag{4.41}$$

Transistor $Mo$ has no fixed operation point, because it is working as a switch: it is therefore not possible to determine its small signal gain since this circuit has a variable operating point. As long as $Mo$ is working in its nonlinear operation mode, there is no small signal gain and consequently no steady-state error: however, when $Mo$ is in saturation (during the slopes), it works as an amplifying stage, and an average small signal gain can roughly be estimated. This is a coarse approximation method, but it serves the purpose to prove that the opamp gain does not have to be large at all. Indeed, simulations confirm that using an NDMOS output transistor and evaluating the voltage gain for different load conditions, the following average small signal gains ($A_{Mo\_DC}$) are obtained:

$$\begin{cases} V_{bat} = 18 \text{ V} \rightarrow |A_{Mo\_DC}| = 40 \text{ dB} \\ V_{bat} = 8 \text{ V} \rightarrow |A_{Mo\_DC}| = 34 \text{ dB} \end{cases} \tag{4.42}$$

Using the results obtained in (4.41) and (4.42) and substituting them in expression (4.40) yields:

$$\begin{cases} |A_{Mo\_DC}| = 40 \text{ dB}, & |T_{signal\_DC}| = 31 \text{ dB} \\ |A_{Mo\_DC}| = 34 \text{ dB}, & |T_{signal\_DC}| = 24 \text{ dB} \end{cases}$$
$$\rightarrow \begin{cases} |A_{opamp\_DC}| = -9 \text{ dB} \\ |A_{opamp\_DC}| = -10 \text{ dB} \end{cases} \tag{4.43}$$

In other words, in order to observe a maximal steady-state error voltage of 0.5 V, the opamp DC voltage gain must be equal to $-9$ dB, which means that the EMI level which is present on the non-inverting opamp input pin is attenuated before reaching the gate of transistor $Mo$. Moreover, knowing that for this particular opamp gain the useful signal range at the non-inverting opamp input differs maximally 0.5 V from the input signal, means that the signal reaching the non-inverting opamp input can be limited to this same level without any loss of signal information.

- **Reducing the dynamic range:** The dynamic range of the signal at the non-inverting opamp input can be reduced according to the steady-state error, and this without any reduction or loss of stability and signal information. On the contrary, limiting the swing at the non-inverting opamp node ensures that the opamp stays in its wanted operation mode, whether or not EMI is present. However, this approach requires a perfect high impedance copy of the signal on the LIN bus. A possible implementation is depicted in Fig. 4.25. The feedback loop which connected the non-inverting opamp input to the LIN bus has been rerouted through a current copying circuit composed of two current mirrors. This circuit makes a high impedance copy of the signal on the LIN bus and feeds it to the non-inverting opamp input to make the circle round. Transistors $M_6$ and $M_7$ further clip this signal between predefined levels, namely between $V_{in} + |V_{GS6}|$ and $V_{in} - |V_{GS7}|$. This way, the EMI signal components in the output signal are firmly contained before reaching the opamp. Furthermore, the opamp has a low gain, which attenuates the disturbance even more before it reaches the sensitive gate of $Mo$. When its unity gain frequency is reached, the clipping structure becomes ineffective, and for this reason, a small internal decoupling capacitor is added between the LIN driver output and ground ($C_{i1}$), ensuring that high frequency disturbances are removed from the output signal

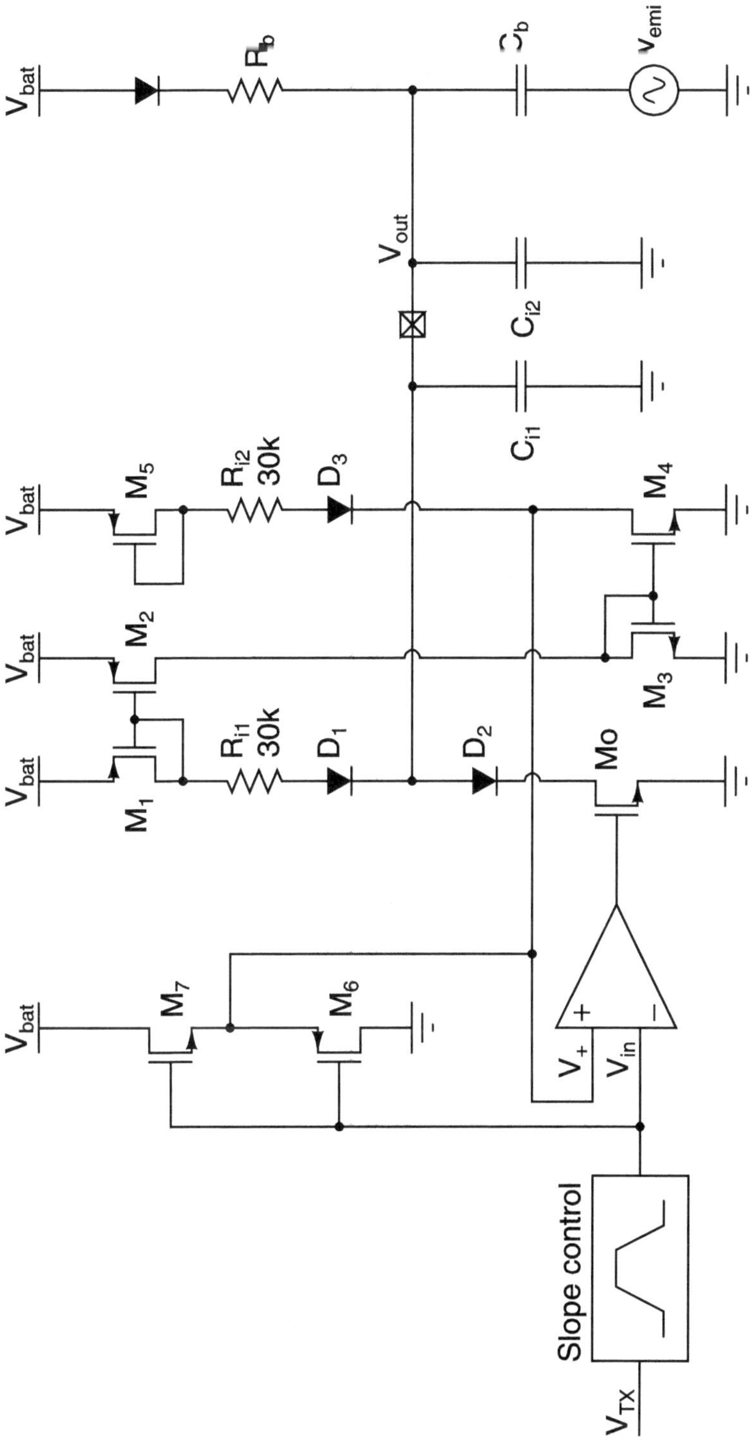


*Figure 4.25.* Final schematic of the proposed LIN driver.

for high frequencies. Capacitor $C_{i2}$ is the external decoupling capacitor, and the sum of $C_{i1}$ and $C_{i2}$ is typically equal to 220 pF, according to the LIN specification [LIN]. Arguably, the current copying structure could be replaced by a resistor, but this would decrease the bandwidth of the signal loop gain function as explained previously, which in turn would invariably deteriorate the stability of this circuit.

### 4.2.2 EMI path 2

- **Introduction:** The parasitic drain-gate capacitance of $Mo$ couples EMI disturbances to the gate of $Mo$: consequently, the output impedance of the opamp has to be quite small in order to decrease this feedthrough as was concluded in Sect. 4.22. In addition, the opamp must cope with input signals which are rail-to-rail, and its gain must be reduced in order to minimize the EMI coupling through *path 1* (depending on the steady-state error, as explained previously). Finally, an extra requirement must be added to this list, namely the output signal of this opamp must not be a linear function of the input signal to prevent distortion by output transistor $Mo$, causing a peculiar form of charge pumping, identified as slope pumping. Each of these substantial constraints are clarified in the following paragraphs.

- **Ideal opamp:** Without any loss of generality, the problem is simplified to its most fundamental issue. Assume that the opamp is ideal, rail-to-rail and further has an output impedance equal to zero, hereby cancelling the EMI feedthrough attributed to *path 2*: observe that, like all ideal opamps, it produces a linear output signal, because the opamp gain ($A_v$) is ideally independent of the magnitude of the input signal. In the event that the output voltage is in a high or low state, the opamp is respectively on and off, driving the output transistor $Mo$ as a switch, as was explained previously. Provided that the parasitic coupling to the gate of $Mo$ through *path 2* and *path 1* is limited, no EMC problems are expected to happen in that case. Let's now look at what happens during the slopes. As long as there is no disturbance coupling into the LIN bus, the gate voltage of $Mo$ is determined by the signal DC open loop gain (and the resulting steady-state error). For a fixed input voltage, this voltage can be expressed as a DC voltage $V_{GSo}$. In this case, provided that $Mo$ is working in saturation, which is the case during the slopes, the drain current of $Mo$ is equal to its mean value as described in (4.37). Consider the presence of a sinusoidal EMI source. As was explained in Sect. 4.2.1, a small EMI disturbance appears at the non-inverting opamp input unscathed ($v_{inp}$), while high disturbances are limited by the clipping structure (composed by $M_6$ and $M_7$ in Fig. 4.25), hereby appearing at the non-inverting opamp input as a square wave. Let's first derive what happens in the event that the EMI signal is small. In that case, the total gate voltage of $Mo$ equals $V_{GSo} + A_v v_{inp}$, where $A_v$ reresents the voltage gain of the ideal, linear opamp. The drain current of $Mo$ ($I_{do}$) is expressed as:

$$I_{do} = \frac{\mu C_{ox}}{2} \cdot \frac{W_o}{L_o} \cdot (V_{GSo} + A_v v_{inp} - V_t)^2 \tag{4.44}$$

This equation can be further expanded as in (4.8), yielding a similar expression for the mean value of the drain current of $Mo$ over time ($\overline{I_{do}}$):

$$\begin{aligned}\overline{I_{do}} &= \frac{\mu C_{ox}}{2} \cdot \frac{W_o}{L_o} \cdot (V_{GSo} - V_t)^2 + \overline{\frac{\mu C_{ox}}{2} \frac{W_o}{L_o} (A_v v_{inp})^2} \\ &= I_{do\_DC} + \frac{\mu C_{ox}}{2} \cdot \frac{W_o}{L_o} \cdot \overline{(A_v v_{inp})^2}\end{aligned} \tag{4.45}$$

In analogy with the charge pumping introduced in the previous chapter, this effect is called slope pumping, since the slope of the output signal is pumped to a different value owing to the nonlinear distortion introduced by $Mo$. Observe that the slope pumping phenomenon occurs during the slope itself, hence its name: once the input signal (labeled $V_{TX}$ in Fig. 4.25) reaches the high or the low state, the opamp drives output transistor $Mo$ to work as a switch, hereby forcing the output to the same level as the input signal. If the slope pumping effect is sufficiently high, the duty cycle is substantially altered, which impacts the correctness of the data transmission. Observe that slope pumping increases the rise time and shortens the fall time: as a consequence, the propagation delay of the rising edge increases while the propagation delay of the falling edge decreases. In addition, the duty cycle is reduced. These effects are illustrated in Fig. 4.26. Slope pumping can be remedied by using an opamp with a nonlinear gain, in particular an opamp whose gain represents the inverse function of the nonlinear distortion introduced by $Mo$. Next paragraph describes how this can be realized in a practical way.

- **Nonlinear rail-to-rail opamp:** A possible solution is shown in Fig. 4.27a. From a small signal point of view, the opamp output impedance is equal to:

$$Z_{out} = \frac{1}{g_{m3}} \tag{4.46}$$

Recall that the output impedance of the opamp is critical in order to minimize the EMI coupling through *path 2*. Furthermore, transistor pair $M_3$-$Mo$ can be considered as a current mirror, which will reduce the slope pumping, as illustrated here below. The drain current of transistor $M_1$ is

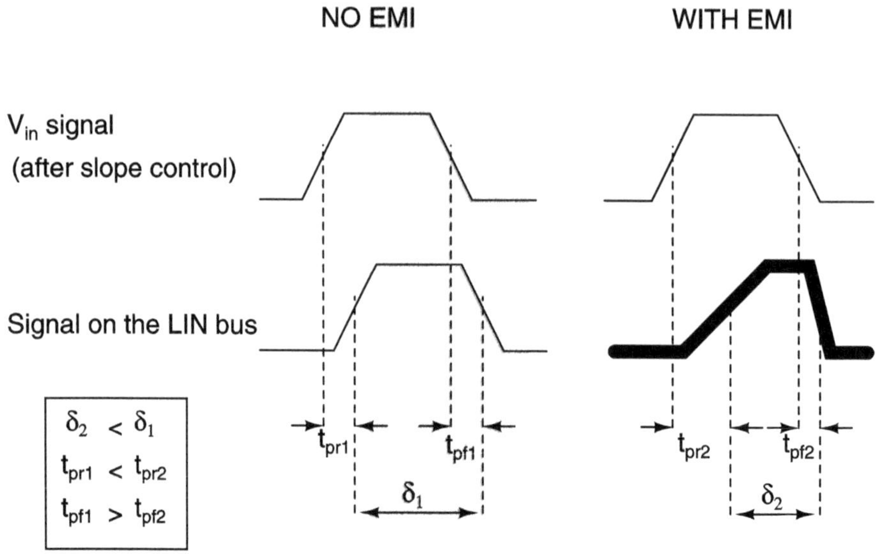


*Figure 4.26.* Effect of slope pumping on the duty cycle, on the rise time and on the fall time.

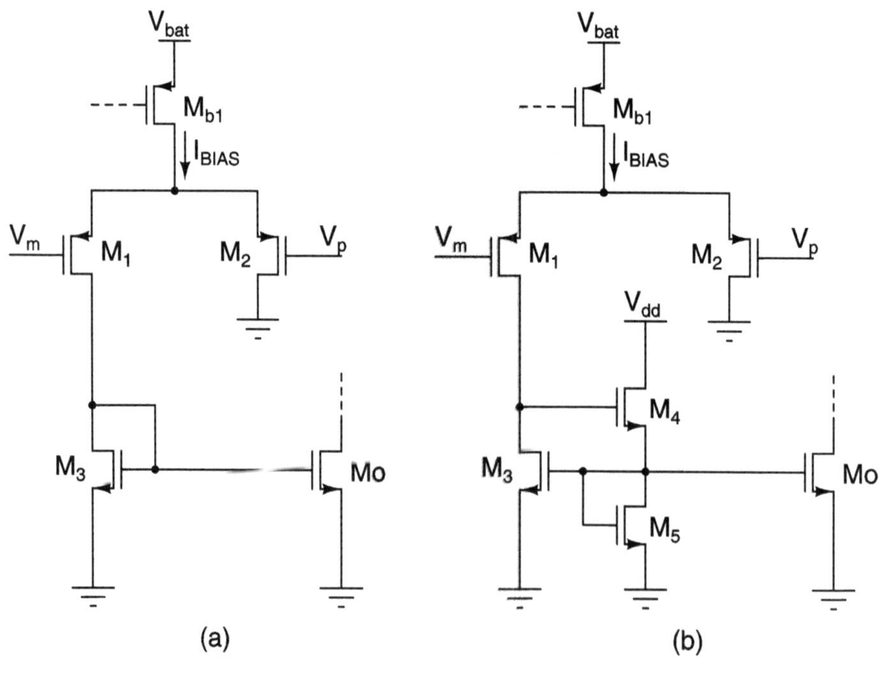


*Figure 4.27.* (a) Basic rail-to-rail opamp structure with low output impedance and low gain. (b) Enhanced rail-to-rail opamp circuit with local feedback to decrease the output impedance.

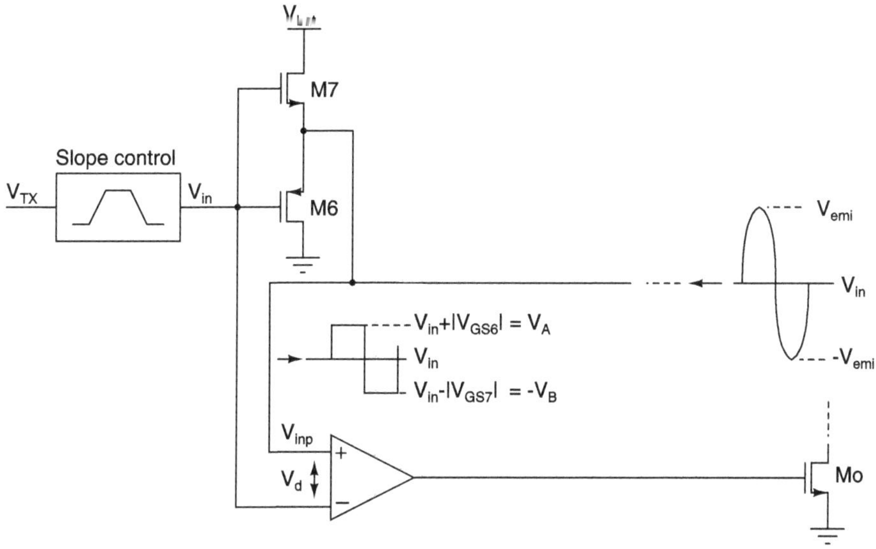


*Figure 4.28.* Slope pumping: clipping of large EMI signals.

expressed as follows:

$$I_{d1} = \frac{\mu C_{ox}}{2} \cdot \frac{W_1}{L_1} \cdot (V_{GS1} + v_{inp} - V_t)^2 \tag{4.47}$$

The average drain current of $Mo$ is consequently equal to:

$$\begin{aligned}\overline{I_{do}} &= \frac{\mu C_{ox}}{2} \cdot \frac{W_o/L_o}{W_3/L_3} \cdot \frac{W_1}{L_1} \cdot (V_{GS1} - V_t)^2 + \frac{\mu C_{ox}}{2} \cdot \frac{W_o/L_o}{W_3/L_3} \cdot \frac{W_1}{L_1} \cdot \overline{v_{inp}^2} \\ &= I_{do\_DC} + \frac{\mu C_{ox}}{2} \cdot \frac{W_o/L_o}{W_3/L_3} \cdot \frac{W_1}{L_1} \cdot \overline{v_{inp}^2}\end{aligned} \tag{4.48}$$

Clearly, comparing previous expression to the one obtained using a linear opamp in (4.45), the EMI induced DC shift of $I_{do}$ can be somewhat reduced by using the opamp depicted in Fig. 4.27a, and by minimizing ratio $\frac{W_1/L_1}{W_3/L_3}$. As can be seen in (4.48), this reduces the appearing slope pumping. However, the major advantage of using this opamp structure resides in its low output impedance. Let's now look at what happens for large EMI signals. Consider that the LIN bus signal $V_{LIN}$ is appropriately copied to the non-inverting opamp input: the (large) EMI which is superposed on that output signal is clipped between $V_{in} + |V_{GS6}|$ and $V_{in} - |V_{GS7}|$ by respectively $M_6$ and $M_7$ (refer to Fig. 4.28). Owing to this, the differential signal between the opamp inputs ($V_d = V_{inp} - V_{inm}$) can roughly be approximated by a square wave with $V_A$ as upper and $-V_B$ as lower level.

As long as its duty cycle is close to 50%, the average drain current of $Mo$ over time can be calculated as previously:

$$\overline{I_{do}} = \lim_{T\to\infty} \frac{1}{T} \int_{-T/2}^{T/2} I_{do}(t) \cdot dt = K \cdot \frac{\mu C_{ox}}{2} \cdot \frac{W_1}{L_1} \cdot \overline{(V_{GS1} + V_d - V_t)^2} \tag{4.49}$$

Where $K$ represents the scale factor of current mirror $M_3$-$Mo$ ($K = \frac{W_o/L_o}{W_3/L_3}$). Above equation can be expanded as follows:

$$\overline{I_{do}} = K \cdot \frac{\mu C_{ox}}{2} \cdot \frac{W_1}{L_1} \cdot ((V_{GS1} - V_t)^2 + 2 \cdot (V_{GS1} - V_t) \cdot \overline{V_d} + \overline{V_d^2}) \tag{4.50}$$

Considering that $V_d$ is a square wave with $V_A$ as upper and $-V_B$ as lower level yields the following mean values:

$$\begin{cases} \overline{V_d} = \frac{V_A - V_B}{2} \\ \overline{V_d^2} = \frac{V_A^2 + V_B^2}{2} \end{cases} \tag{4.51}$$

Substituting (4.51) in (4.50) gives:

$$\overline{I_{do}} = \frac{K \cdot I_{BIAS}}{2} + \frac{K \cdot I_{BIAS}}{V_{GS1} - V_t} \cdot \left( \frac{V_A - V_B}{2} \right) + \frac{K \cdot \mu C_{ox}}{2} \cdot \frac{W_1}{L_1} \cdot \left( \frac{{V_A}^2 + {V_B}^2}{2} \right) \tag{4.52}$$

Where $I_{BIAS}$ represents the DC tail current of the differential pair. Previous expression represents the slope pumping appearing when large EMI signals are injected in the LIN bus, and limited accordingly by the clipping transistors $M_6$ and $M_7$. As long as $V_A$ and $V_B$ are small, the third term in (4.52) may be neglected, yielding the same expression as in [Red07a]:

$$\overline{I_{do}} \approx K \cdot \frac{I_{BIAS}}{2} + K \cdot \frac{I_{BIAS}}{V_{GS1} - V_t} \cdot \left( \frac{V_A - V_B}{2} \right) \tag{4.53}$$

This means that ideally, $V_A$ should be equal to $V_B$ in order to have the same average output current whether or not EMI is present (and equivalently, the same mean output voltage). However, assuming that the clipping transistors are well matched so as to clip at an equal level (ensuring that $V_A = V_B$), this will only be true as long as the steady-state error is zero. In practice, the steady-state error is never zero and depending on the total bus load it even varies substantially, so that a certain amount of slope pumping is inevitable. The following paragraph examines how slope pumping caused by the steady-state error impacts the correct operation of the LIN driver, and how it can be reduced even further.

- **Steady-state error related to slope pumping:** Calculating the maximal amount of slope pumping, as well as its exact impact on the duty cycle, is difficult to perform accurately. As said previously, the slope pumping effect is highly affected by the steady-state error, which in turn varies with the load. Furthermore, (4.53) is a linearized formula, which is relevant as long as $V_A$ and $V_B$ remain below the overdrive voltage of $M_1$ ($V_{GS1} - V_t$). Finally, it has been assumed so far that the clipped EMI signal is a perfect square wave, which is a very rough approximation depending on the behavior of the clipping transistors $M_6$ and $M_7$, as well as on the magnitude of the EMI signal, which is directly responsible for the steepness of the square wave edges. However, the focus of this paragraph lies not on obtaining a precise figure, but on checking the order of magnitude of the resulting duty cycle deviation, and on identifying which parameters reduce the impact of slope pumping. To this end, expression (4.53) is rewritten as a function of the steady-state error $V_{SSE}$, by replacing $V_A$ with $V_{in} + V_{SSE}$ and $V_B$ with $V_{in} - V_{SSE}$. This yields:

$$\overline{I_d} \approx K \cdot \frac{I_{BIAS}}{2} + K \cdot \frac{I_{BIAS}}{V_{GS1} - V_t} \cdot V_{SSE} \tag{4.54}$$

As specified earlier in Sect. 4.2.1, the steady-state error voltage is maximally equal to 0.5 V. Designing for an overdrive voltage $V_{GS1} - V_t$ of 1 V, yields:

$$\overline{I_d} \approx K \cdot I_{BIAS} \tag{4.55}$$

This means that the average drain current is doubled owing to slope pumping. Knowing that the slope time is equal to 5 μs, and measuring the duty cycle between the 50% levels of the battery voltage, signifies that the original duty cycle of 50% is reduced to a theoretical 42.5% at a maximal transmission speed of 20 kbit/s owing to slope pumping. This worst case figure lies barely within the specification limitation, which specifies that the duty cycle should be comprised between 41.7% and 59% at a transmission speed of 10.4 kbit/s, and between 39.6% and 58.1% at a transmission speed of 20 kbit/s at all times [LIN]. The resulting safety margin is small, but can be raised by increasing the overdrive voltage $V_{GS1} - V_t$ as is apparent from (4.54). Another possible approach to decrease the impact of slope pumping is to lengthen the duration of the duty cycle: indeed, slope pumping shortens the duty cycle, and since the specification provides a lower boundary which is almost as high as the upper boundary, the original duty cycle could be boosted to a safer nominal value of e.g. 55%: this would require two different slope control blocks treating the rising and the falling slope in a different way, but could as well be implemented digitally in the processor generating the original transmit signal. However, this evidently increases the duty cycle when no EMI is present on the LIN bus.

This potential improvement has not been researched further. Finally, slope pumping can be reduced by decreasing the opamp gain, but this increases the steady-state error. This approach has been succesfully used in the second LIN driver design, and is covered in full in Sect. 4.3.2.

- **Enhanced nonlinear rail-to-rail opamp:** As derived previously, the opamp output impedance must be as small as possible in order to minimize the EMI coupling through the parasitic drain-gate capacitance of output transistor $Mo$ (*path* 2). This can be realized by implementing a local feedback structure, which is effectively used in current mirrors in order to make them more resisting to EMI [Red05, Red06a]. This concept is shown in Fig. 4.27b. Transistor $M_4$ isolates the sensitive gate of $Mo$ from the drain of $M_3$, while $M_5$ completes the biasing. The purpose of transistors $M_4$ and $M_5$ is to keep the gate of $Mo$ at a fixed level using negative feedback. Looking from a small signal point of view, the output impedance of the opamp is now equal to:

$$Z_{out} = \frac{1}{g_{m4} + g_{m5} + r_o g_{m3} g_{m4}} \tag{4.56}$$

  where $r_o$ is the parallel combination of the (high) output resistances of $M_1$ and $M_3$. Expression (4.56) illustrates that the output impedance of the opamp can be tremendously reduced using local feedback, while still preserving the necessary nonlinear gain.

- **Output impedance asymmetry:** The conclusion of previous subsection is correct when transistors $M_3$, $M_4$, $M_5$ are biased in their saturation region: however, the opamp (and the complete circuit for that matter) is operated in its nonlinear mode, and in this case its output is either high, either low, in which case a small signal analysis does not apply any longer. Let's consider what happens in this event.

  - Firstly, if the opamp output is high, transistors $M_3$, $M_4$ and $M_5$ are active. The opamp output impedance is then indeed very low as was confirmed previously, and $M_4$ sources the current which flows through $M_5$ and which charges the parasitic gate-source capacitance of $Mo$ ($C_{gso}$).
  - Secondly, if the opamp output is low, then $M_3$ and $M_4$ are off, while diode connected transistor $M_5$ sinks the charge which is stored in $C_{gso}$. The output impedance of the opamp is therefore much higher, since there is no longer gain to reduce the opamp output impedance with the negative feedback. And for high EMI levels coupling from the drain to the gate of $Mo$, a situation might occur when the charge which is "pumped" into $C_{gso}$ becomes larger than the charge that $M_5$

is capable of sinking. If such a situation happens, the mean voltage on the gate of $Mo$ increases, forcing a higher drain current to flow through $Mo$. This causes the output signal to drop, and may thus result in a distorted output signal. As a consequence, the duty cycle of such output signal is no longer equal to its original value, generating an erroneous data transmission. To prevent excessive charge pumping across $C_{gso}$, the ration $W/L$ of $M_5$ may be increased. However, owing to the nonlinear I-V characteristic of a MOS diode connected transistor, the parasitic $C_{gso}$ is not discharged linearly, which makes this structure inherently susceptible to EMI induced DC shift. In order to be certain that the parasitic $C_{gso}$ has a continuous discharge path, a resistor of 120 Ω connected between the gate of $Mo$ and ground has been added (not drawn on the schematic in Fig. 4.27b). Unfortunately, this in turn, accounts for an increased static power consumption through $M_4$. This issue has been studied and improved in the second LIN driver design.

### 4.2.3 Slope control function

The circuit generating the slope is not disturbed by EMI any longer, since it is now fully separated from the output stage, and can be designed using a current starved inverter directing a constant current through capacitor $C_f$ (Fig. 4.29). Current starved inverters are widely appreciated and very frequently used when designing delay elements [Joh88]. Resistor $R_{bias}$ sets the current flowing through the current starved inverter and makes this current linearly proportional to the battery voltage. This way, the total slope time remains constant for varying battery voltages.

### 4.2.4 Measurements

The proposed LIN driver has been integrated using the AMIS I3T80 0.35 μm high-voltage technology [Moe02]. A microphotograph of the circuit is shown in Fig. 4.30. Conforming to the DPI injection measurements (zone 1, maximal EMI level severity), 5 W forward power was injected in the LIN bus through the bus capacitance, and this for frequencies ranging from 150 kHz up to 1 GHz [DPI]: refer to the measurement setup depicted in Fig. 4.31. Injecting 5 W of forward power requires that the EMI source (50 Ω internal resistance) reaches an amplitude level of maximally 45 V as was summarized in Table 2.2, which constitutes an impressive requirement to fulfill.

While performing these measurements, the voltage on the LIN bus was continuously monitored on an oscilloscope: an example of a resulting output signal is shown in Fig. 4.32. Although the signal on the LIN bus is heavily corrupted by EMI, the duty cycle remains (almost) unchanged. The duty cycle was measured between the 50% levels of the battery voltage and by filtering the signal

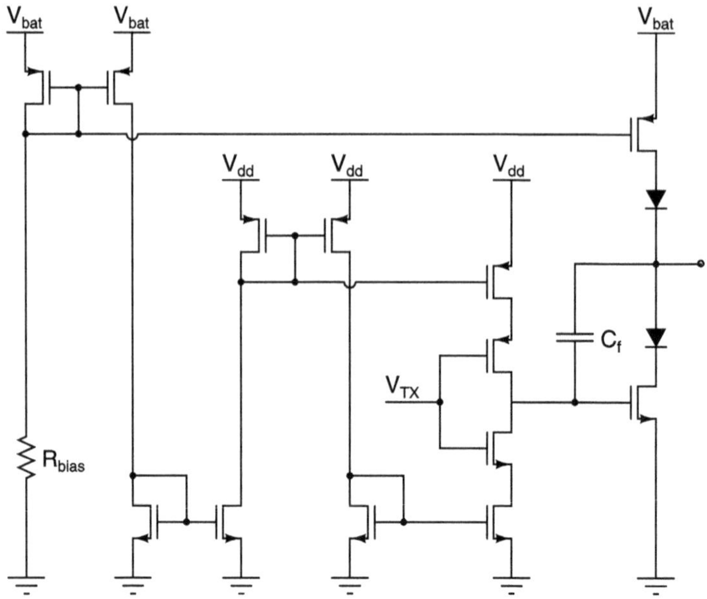


*Figure 4.29.* Slope control circuit.

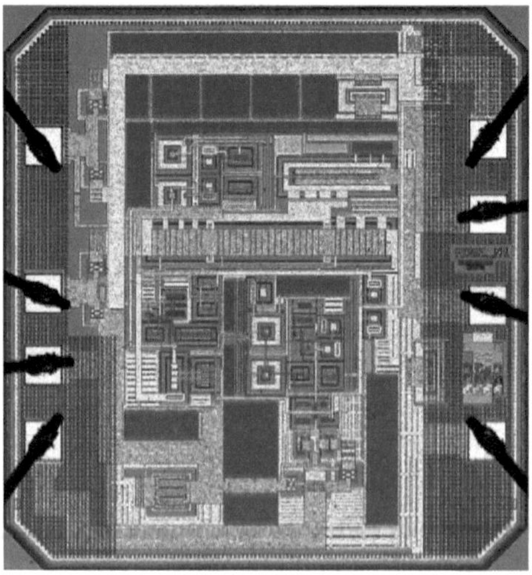

*Figure 4.30.* Microphotograph of the LIN driver. The size of the circuit is approximately 1.2 mm × 1.3 mm.

on the LIN bus numerically in order to get rid of the high frequency components. A listing of the performed measurements, and the resulting duty cycle and propagation delays, is shown in Table 4.3. The largest measured duty cy-

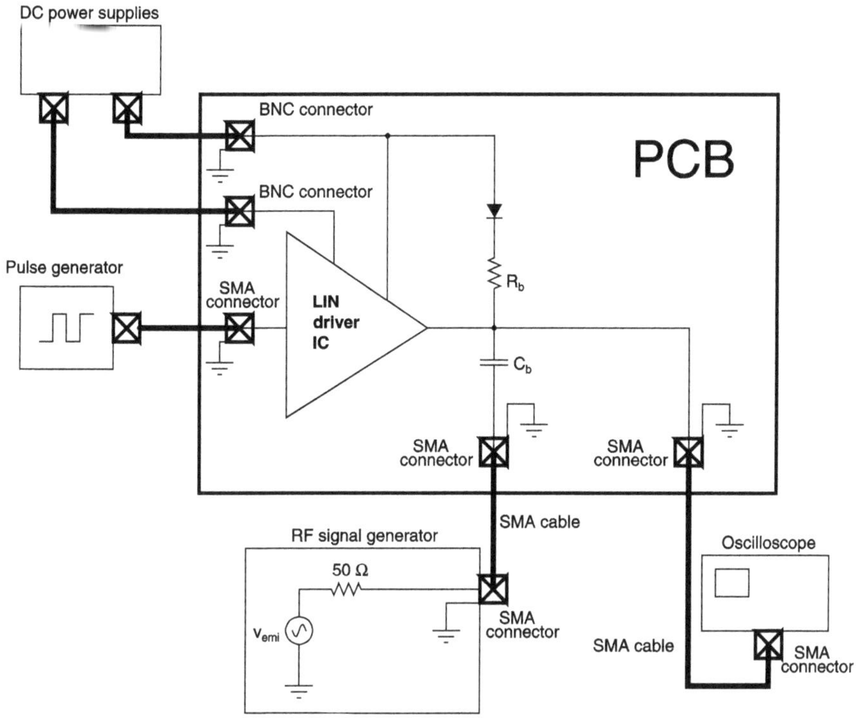


*Figure 4.31.* Direct power injection measurement setup, as described in the DPI specification [DPI].

cle deviation of this filtered waveform is equal to 7.2%, which lies within the range that is specified by the LIN specification, namely the duty cycle should be comprised between 41.7% and 59% at a transmission speed of 10.4 kbit/s, and between 39.6% and 58.1% at a transmission speed of 20 kbit/s [LIN]. Observe that the maximal duty cycle deviation occurs at an EMI frequency of 1 MHz. This is logical, since the bus capacitance represents a non-negligible impedance for the lowest EMI frequencies which reduces the EMI injection on the LIN bus, while for higher frequencies, the internal and external decoupling capacitors (total value equal to 220 pF) drain most of the injected EMI. Finally, observe that this maximal duty cycle deviation (at an EMI frequency of 1.1 MHz) lies very close to the theoretical value derived in Sect. 4.2.2, when discussing the slope pumping phenomenon. Not surprisingly, the duty cycle is almost never perfectly equal to 50% owing to the slope pumping. This slope pumping effect is clearly experienced in the increased propagation delay of the rising slope of the LIN signal, as observed in Table 4.3. The deviation caused by slope pumping is considerable, yet it is still well within the specification limits. However, note that slope pumping is predominantly present, particu-

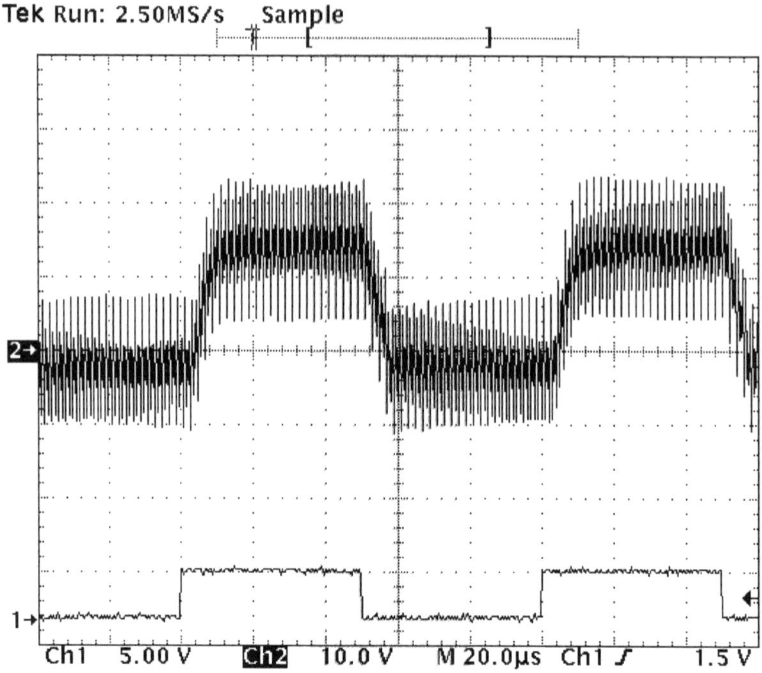


*Figure 4.32.* Example of an EMI distorted output signal: although heavily corrupted, the resulting signal after numerical filtering possesses the same duty cycle as the transmitted signal $V_{TX}$ (at the bottom).

*Table 4.3.* DPI measurement results ($V_{bat} = 15$ V, $V_{dd} = 3.3$ V, $R_b = 1$ kΩ, $C_b = 1$ nF, $V_{TX}$ transmission speed = 10 kbit/s).

| *EMI frequency (MHz)* | *Forward EMI power (W)* | *Duty cycle (%)* | *Prop. delay (rising edge) (μs)* | *Prop. delay (falling edge) (μs)* |
|---|---|---|---|---|
| – | – | 50 | 4.8 | 4.8 |
| 0.3 | 5 | 43.6 | 9.6 | 3.2 |
| 1 | 5 | 42.8 | 10.4 | 3.2 |
| 5 | 5 | 45.6 | 8.4 | 4.0 |
| 10 | 5 | 46.4 | 8.0 | 4.4 |
| 40 | 5.2 | 49.6 | 5.2 | 4.8 |
| 100 | 5.0 | 49.2 | 5.6 | 4.8 |
| 200 | 5.1 | 49.6 | 5.6 | 5.2 |
| 300 | 5.2 | 50.0 | 4.8 | 4.8 |

larly on the rising slope, as predicted in Fig. 4.26. This issue is studied and suitably remedied in the second LIN driver design.

## 4.3 EMI resisting LIN driver topology: LIN driver 2

Throughout Sect. 4.2, an integrated LIN driver has been developed with a high degree of immunity against EMI. There are, however, two major imperfections associated to this design:

- Firstly, the opamp has to have a very low output impedance in order to minimize the EMI coupling through the parasitic drain-gate capacitance of $Mo$ to its gate (labeled *path 2* in Fig. 4.24): this low opamp output impedance was achieved at the severe penalty of a heavy and continuous power consumption in the first LIN driver design, as explained in Sect. 4.2.2.

- Secondly, the EMI which is copied to the non-inverting opamp input and limited between specific levels introduces a considerable shift in the rising slope of $V_{LIN}$: this shift has been identified as slope pumping in Sect. 4.2.2. Slope pumping distorts the duty cycle and the propagation delay of the LIN bus signal.

Both issues are tackled in the second LIN driver design.

### 4.3.1 Smart-power mode

The opamp output impedance must be low to decrease the high frequency coupling from the LIN bus to the gate of $Mo$ through its parasitic drain-gate capacitance: the moment high frequency EMI couples to the gate of $Mo$, it pumps the net charge across the parasitic gate-source capacitance of $Mo$ ($C_{gso}$) upwards, increasing the gate voltage of $Mo$ and hereby progressively and irrevocably shorting the LIN bus to ground. Refer to the opamp schematic used in LIN driver 1 (Fig. 4.27b). In this opamp, $M_4$ and $M_5$ keep $V_o$ at a fixed DC level using local negative feedback, preventing charge pumping: however, the opamp output impedance is highly asymmetric. In order to provide a constant discharge path for the parasitic gate source capacitor $C_{gso}$ of $Mo$, a resistor of 120 Ω, connected between the gate of $Mo$ and the ground, has been added in the first LIN driver design (Fig. 4.33a). Of course, this accounts for a high static power consumption. Considering all these reasons, $M_5$ has been redesigned as a current source, as shown in Fig. 4.33b. This way, the parasitic gate-source capacitor of $Mo$ ($C_{gso}$) is discharged linearly, since $M_5$ operates as a current source as long as $V_o$ stays above $V_{dssat5}$. If $V_{out}$ drops below $V_{dssat5}$, $M_5$ operates in its triode region, shorting $V_o$ forcibly to ground. In order to overcome the high static power consumption, a smart-power circuit has been implemented which switches on and shorts the gate of $Mo$ to its source when the voltage $V_{LIN}$ collapses owing to a large EMI which is injected into the LIN bus. This allows the continuous bias current through $M_4$ and $M_5$ to be small and barely sufficient to pull the gate of $Mo$ to ground in the absence of EMI,

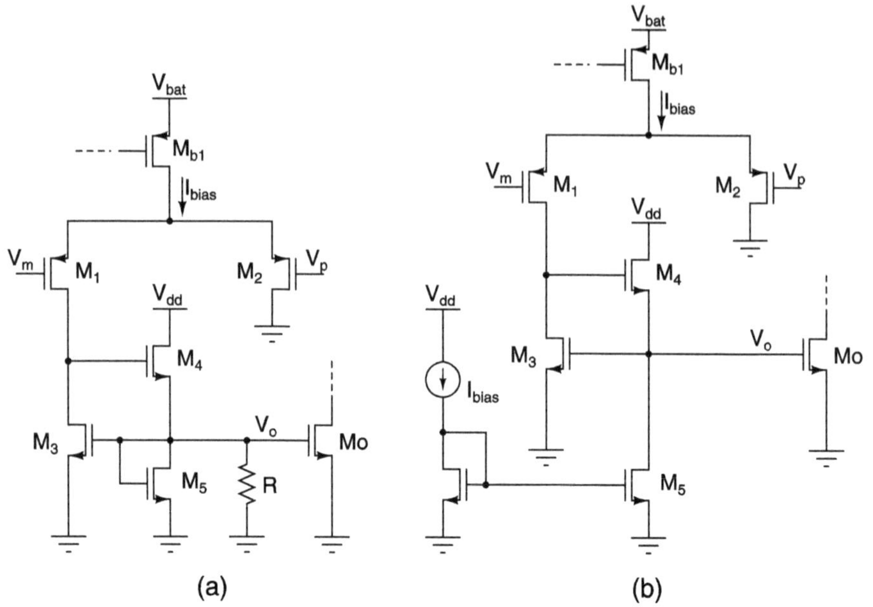


*Figure 4.33.* (a) Opamp with diode connected $M_5$ (as in LIN driver 1). (b) Opamp biased with a (small) constant bias current through $M_5$ (as in LIN driver 2).

which minimizes and benefits the overall power consumption (the continuous bias current through $M_4$ has been chosen equal to 2.3 mA in this design).

The smart-power capability in question has been implemented as follows: refer to Fig. 4.34. The clipping structure detects when the LIN bus output collapses to ground owing to large EMI levels: $M_7$ detects the sudden signal drop at the non-inverting opamp input, and turns on $M_8$. This current is copied, scaled and sinked via $M_{11}$ hereby pulling the gate of $Mo$ to ground, which restores $V_{LIN}$. As soon as the necessary output level is reached, or as soon as the EMI injection stops, $M_{11}$ shuts off and the LIN driver resumes its low power operation. The average opamp power consumption for varying EMI injected levels is enumerated in Table 4.4. Observe that the average consumed power in the present LIN driver opamp is lower than in LIN driver 1. More importantly, observe that the present LIN driver opamp switches to a higher power consumption for increasing EMI levels. This indicates that the smart-power principle is working properly, and that a much smaller continuous bias current through $M_5$ may be used, reducing the static power consumption to a much lesser value than indicated in Table 4.4.

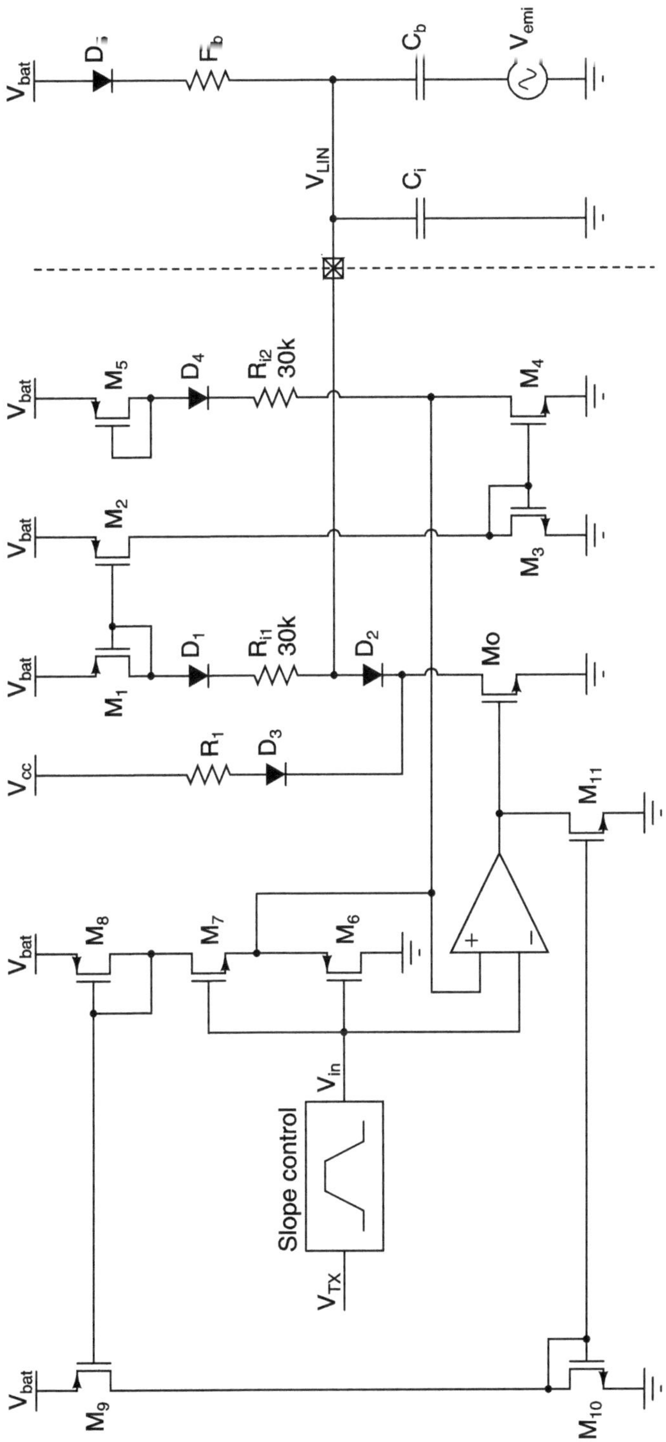


*Figure 4.34.* EMI resisting smart-power LIN driver with reduced slope pumping.

*Table 4.4.* Opamp average power consumption ($V_{bat} = 12$ V, $V_{dd} = 3.3$ V, $R_b = 1$ kΩ, $C_b =$ 1 nF, $V_{TX}$ transmission speed = 20 kbit/s).

| *EMI* | *Opamp $P_{average}$ LIN driver 1 (mW)* | *Opamp $P_{average}$ LIN driver 2 (mW)* |
|---|---|---|
| no EMI | 28 | 11 |
| EMI = 10 V@1 MHz | 26 | 12 |
| EMI = 45 V@1 MHz | 28 | 14 |
| EMI = 10 V@10 MHz | 26 | 12 |
| EMI = 45 V@10 MHz | 28 | 15 |
| EMI = 10 V@100 MHz | 26 | 11 |
| EMI = 45 V@100 MHz | 25 | 13 |

### 4.3.2 Slope pumping reduction

Slope pumping distorts the rising and falling edges of the LIN bus signal and consequently corrupts the duty cycle and the propagation delay, as has been illustrated in Sect. 4.2.2. According to (4.52), the average value of the drain current flowing through $Mo$ is pumped to a higher value owing to the large EMI which is superposed on the non-inverting opamp input. Observe that the slope pumping phenomenon solely occurs during slopes: once $V_{LIN}$ reaches the high or the low state, the opamp drives output transistor $Mo$ as a switch, hereby forcing the output to the same level as input signal $V_{in}$. As can be seen in expression (4.52) and as has been illustrated previously in Fig. 4.26, slope pumping lengthens the rise time while shorting the fall time of $V_{LIN}$: clearly, the latter is not an issue, but the former severely distorts the duty cycle and perturbs the propagation delay of the LIN signal's rising edge. It is therefore important to reduce the slope pumping on the rising slope as much as possible. The slope pumping issue has already been discussed anteriorly, albeit using the simplified linearized expression derived in (4.53). This section weighs the full impact of slope pumping, and considers how a reasonable slope pumping reduction can be achieved. As can be seen in (4.52), there are basically three ways to achieve a slope pumping reduction:

- **Decrease $V_A$ and $V_B$.** The first possibility is to decrease $V_A$ and $V_B$. However, both have to be larger than the steady-state error. As calculated in Sect. 4.2.1, the steady-state error is equal to 0.5 V, which defines the lower boundary of $V_A$ and $V_B$. In this case, $V_A$ and $V_B$ are defined by clipping transistors $M_6$ and $M_7$, and are respectively equal to $V_{in} + |V_{GS6}|$ and $V_{in} - |V_{GS7}|$ (refer to Fig. 4.34).

- **Ensure that $V_A = V_B$.** If $V_A$ is perfectly equal to $V_B$, the second (linear) term in (4.52) is canceled, which reduces the slope pumping. Unfortunately, $V_A$ and $V_B$ are directly affected by the steady-state error, which

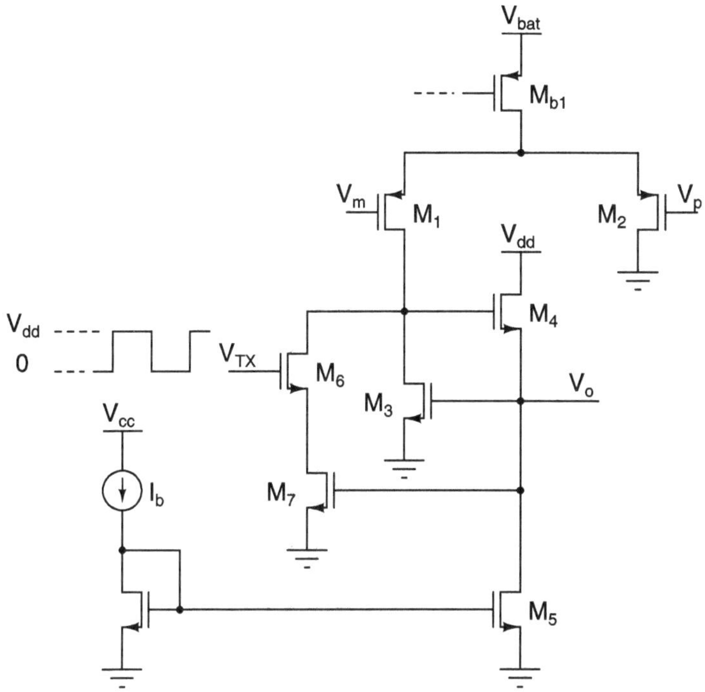


*Figure 4.35.* Opamp biased with a (small) constant bias current. Transistor $M_6$ is switched on when $V_{TX}$ toggles to a high state, hereby lowering the voltage gain of the opamp and attenuating the slope pumping during the rising edge.

in turn depends on the load present on the LIN bus: moreover, this load is not a priori known. This varying steady-state error causes $V_A$ and $V_B$ to constantly differ from each other. Moreover, the quadratic term in (4.52) remains unaffected by this approach.

- **Decrease $K$.** A third way to decrease slope pumping is to reduce the current mirror scaling factor $K$. This corresponds to decreasing the opamp gain: however, this equally increases the steady-state error and limits the maximum voltage level at the opamp output, which in turn increases the low state of the LIN bus output signal (refer to Sect. 4.2.1). It is therefore important to reduce the opamp gain solely during the rising slopes and the high LIN bus output state, while reverting to a high opamp gain during the low output state. This has been implemented as follows (Fig. 4.35): $M_6$ turns on as soon as the digital input signal ($V_{TX}$) is high. Transistor $M_7$ is then in parallel with $M_3$, hereby reducing the opamp gain. Once $V_{TX}$ drops, $M_6$ shuts off, the opamp output signal increases and forces a low output state on the LIN bus. This way, slope pumping in the rising edge is effectively countered.

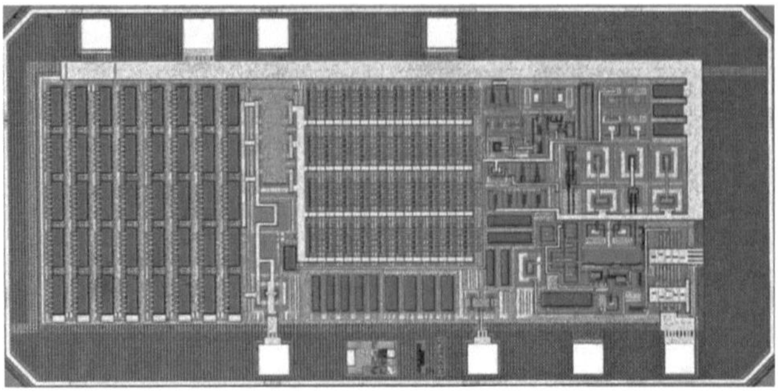

*Figure 4.36.* Microphotograph of the EMI resisting smart-power LIN driver with reduced slope pumping (LIN driver 2). The size of the circuit is approximately 2.1 mm × 1.1 mm.

*Table 4.5.* DPI measurement results ($V_{bat} = 12$ V, $V_{cc} = 3.3$ V, $R_b = 1$ kΩ, $C_b = 1$ nF, $V_{TX}$ transmission speed = 20 kbit/s).

| *EMI frequency (MHz)* | *Forward EMI power (W)* | *Duty cycle (%)* | *Prop. delay (rising edge) (μs)* | *Prop. delay (falling edge) (μs)* |
|---|---|---|---|---|
| – | – | 49.2 | 3 | 3 |
| 0.3 | 5.99 | 42.4 | 4.4 | 1 |
| 1 | 6.02 | 42.8 | 4.2 | 1 |
| 10 | 5.95 | 44 | 3.6 | 1 |
| 40 | 5.25 | 45.6 | 3.2 | 1.4 |
| 100 | 5.26 | 46 | 3 | 1.4 |
| 500 | 5.01 | 46 | 2.6 | 1 |
| 1000 | 5.05 | 49.2 | 3 | 3 |

### 4.3.3 Measurements

The proposed LIN driver (Fig. 4.35) has been integrated using the AMIS I3T80 high-voltage 0.35 μm CMOS technology: a microphotograph of the circuit is shown in Fig. 4.36 [Moe02]. A forward EMI power of maximally 5 W has been injected through the external bus capacitance $C_b$, ranging from 150 kHz to 1 GHz according to the DPI specification [DPI]: this corresponds to an EMI source amplitude of 45 V, as mentioned previously. Measurements show that the proposed LIN driver withstands this 5 W DPI test for the listed frequency range (Fig. 4.37): Table 4.5 summarizes the results. Observe the almost constant propagation delay compared to the ones enumerated in Table 4.3 for LIN driver 1, as well as the minimal impact of EMI on the duty cycle. As specified in [LIN], the duty cycle is comprised between 39.6% and 58.1% at all times for a transmission speed of 20 kbit/s.

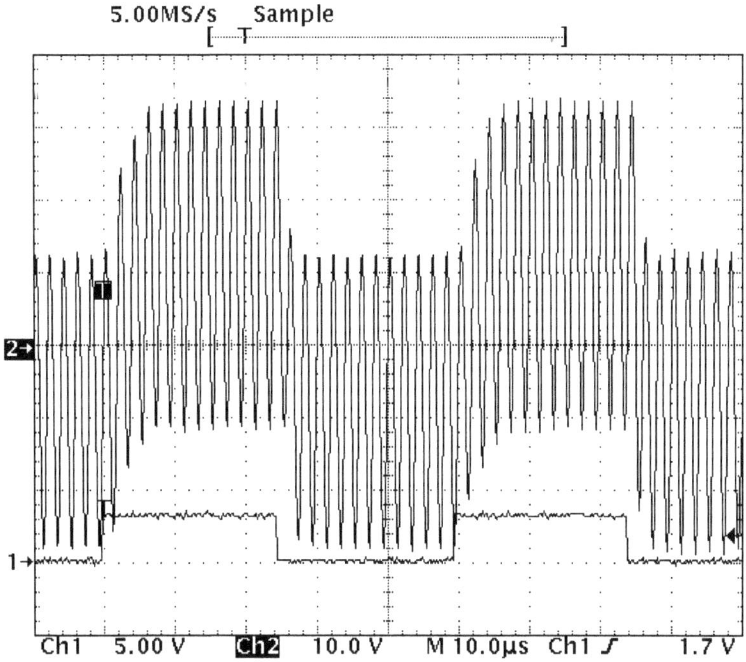


*Figure 4.37.* 5 W DPI measurement of the EMI resisting smart-power LIN driver: although the LIN bus is heavily distorted, the duty cycle is not altered.

Chapter 5

# EMI Resisting Analog Input Circuits

*Mobile phone makers and car manufacturers are investigating claims that handsets can cause car safety airbags to inflate and interfere with automatic braking systems. Tests carried out by Volvo in Sweden found that phones operating independently of car electrics can trigger airbags and interfere momentarily with control systems.*

—Quoted from [Arm07]

## 1 Introduction

Analog integrated circuits are as a rule very susceptible to EMI which is conveyed into their input terminals. Unlike digital circuits, which exhibit an intrinsic robustness against small input signal variations depending on their threshold levels, analog circuitry treats an EMI disturbance applied at (one of) its input terminals as an ordinary input signal. These interferences may be in-band or out-of-band. If in-band, they are no longer distinguishable from wanted signals, unless extra information concerning the latter is available. In order to differentiate EMI signals from wanted signals, differential signaling may be used, since disturbances injected on neighboring PCB tracks or connecting wires tend to manifest themselves as common-mode signals at the IC inputs [1]. This improves the electromagnetic immunity of the circuit, but poses serious constraints on its common-mode rejection ratio (CMRR). If out-of-band, EMI signals can theoretically be filtered or separated from the wanted input signal(s). This is however easier said than done, since careless or inappropriate filtering may cause nonlinear distortion and even trigger the very detrimental

[1] The common-mode input signal comprises the signal component which is equal for both inputs ($v_{i\_cm} = (v_{i1} + v_{i2})/2$), while the differential-mode input signal identifies the signal component which appears between both input pins ($v_{i\_dm} = v_{i1} - v_{i2}$) [Gra01].

J.-M. Redouté, M. Steyaert, *EMC of Analog Integrated Circuits,*
Analog Circuits and Signal Processing,
DOI 10.1007/978-90-481-3230-0_5, 

DC shift phenomenon, as was shown in Chap. 3. In addition, the circuit itself may rectify high frequency EMI signals that are situated well above its working bandwidth, generating undesirable harmonic components, troublesome intermodulation products and an inevitable DC shift. In this chapter, the EMI susceptibility of analog input circuits is studied by means of two case studies:

- **Case study 1: improving the electromagnetic immunity of CMOS operational amplifiers.** The first case study describes the EMI susceptibility of operational amplifiers, when EMI is conveyed into their input terminals. Operational amplifiers are one of the most popular analog building blocks employed in the design of analog and mixed-signal IC's, but, unfortunately, they are in general extremely sensitive to EMI which is conveyed to their input terminals. In particular, operational amplifiers are disturbed by a DC offset which is induced by EMI distorting the differential input stage, owing to an asymmetric slew rate as well as to the nonlinear behavior of the differential input stage.

- **Case study 2: an instrumentation amplifier input circuit with a high rejection to common-mode EMI.** The second case study analyzes the EMI susceptibility of an instrumentation amplifier when very large EMI common-mode signals are injected into its input terminals. Because the EMI common-mode signals can be much larger than the wanted differential signal level, the slightest mismatch translates a portion of the common-mode disturbance into a differential-mode signal owing to the finite CMRR. This is of course highly undesirable, since this EMI induced differential signal may mix and intermodulate with the wanted differential signal or disturb subsequent circuits. This case study describes how the matching of the input protection stage may be improved, in order to increase the common-mode rejection ratio.

## 2 Case study 1: electromagnetic immunity of CMOS operational amplifiers

CMOS operational amplifiers are one of the most prevalent analog building blocks but, unfortunately, they are also extremely sensitive to EMI which is injected into their terminals. Many articles and papers have been published on this subject, and many measurements and calculations comparing the EMI behavior of various opamp topologies have been conducted and reported [Gra97]. EMI measurements comparing the EMI induced DC output offset of many commercially available opamps have been presented in [Pou94]. Operational amplifiers may rectify EMI signals lying at frequencies which are situated well above their bandwidth, which typically results in harmonics, intermodulation products and DC shift. These effects have been discussed in Chap. 3, where it has been illustrated that especially DC shift acts very detrimentally on the behavior of analog circuits since it alters the correct DC biasing, hereby even possibly debiasing the full circuit completely. The black box abstraction of an operational amplifier contains output terminals, input terminals and a power supply. This signifies that EMI may affect an operational amplifier in three distinct ways:

- **EMI injection in the power supply:** in general, the performance of a system influenced by power supply variations is described by the power supply rejection ratio (PSRR), which is the ratio of the signal transfer function to the power supply transfer function to the output node. Consequently, the PSRR of operational amplifiers can be improved by increasing the gain bandwidth product (GBW) of the amplifier in question [Ste91]. However, the PSRR is in practice always limited by mismatches in the circuit, even for improved schematics [Gie93]. In addition, at high EMI frequencies, a portion of the EMI injected in the power supply terminal interferes with the opamp, and hereby partly impairs or even fully disrupts its correct operation. Internal and external filtering prevents EMI from coupling into the opamp through the power supply, and hereby generating DC shift. This is illustrated amongst others in [Kov00], where the behavior of 741 opamps from three different manufacturers has been compared with measurements when the same EMI is injected into their respective power supplies. The major conclusion stemming from this study revealed that a small external $R$-$C$ low-pass filter placed close to the supply pins ($R = 47\ \Omega$, $C = 10\ \mu\text{F}$) was sufficient to significantly filter the EMI, since it reduced the DC shift on the output of the opamp from 5.7 V to barely 300 mV. This comparison illustrates once again the importance of decoupling the power supply lines. Not surprisingly, the failures induced in opamps when EMI is injected in the power supply rails has been linked to their power supply rejection ratio [Spe99].

- **EMI injection in the output node:** the effect of EMI injected in output circuits has been studied extensively in Chap. 4. Opamp outputs are either common-source, either common-drain structures, and depending on the used topology, care must be ensured in each particular case in order to prevent EMI from generating DC shift in the output stage. For instance, a source follower output (e.g. push-pull type) provides a low output impedance, but is very susceptible to EMI which is injected into its output terminals. Conversely, a common-source output of a two stage Miller opamp couples EMI into the opamp circuit through the Miller compensation capacitor, hereby disturbing the output driving transistor(s) comparable to what has been observed in the classic LIN driver, which was described in Chap. 4.

- **EMI injection in the input pins:** in contrast with the two previous cases, the EMI signals which are superposed on the opamp inputs are by far the ones that are the most intricate to counter and filter correctly. Indeed, the wanted input signals are usually quite small, and using external filtering modifies and attenuates them even more. In addition, extra protection circuits often generate extra input noise (e.g. when using a $R$-$C$ low-pass filter). The EMI susceptibility of operational amplifier input stages is related to three basic phenomena [Cor05]. These are studied and discussed in detail in the following paragraphs, and are summarized as follows:

  - **Slew rate asymmetry:** this effect is probably the first one that has been cited in the study of electromagnetic susceptibility of operational amplifier input circuits. As illustrated further on, asymmetry in the slew rate leads to DC shift.

  - **Effect of strong nonlinear behavior of the input stage:** this effect is often referred to as the "effect of parasitic capacitances" in literature [Gra97]. This terminology is, however, particularly confusing, since parasitic capacitances influence the slew rate, as well as the strong and weak nonlinear distortions taking place in the input differential pair stage. When the input differential stage is overdriven by a very large high frequency EMI signal, the input transistors are alternately forced into cut-off, generating strong nonlinear distortions and a subsequent DC shift. This effect can be reduced by minimizing the parasitic capacitances of the input differential pair.

  - **Effect of weak nonlinear behavior of the input stage:** this is without doubt, the most important and intricate EMI effect which is responsible for disrupting the performance of the differential input stage. When a high frequency sinusoidal EMI signal that does not force the amplifier into cut-off is conveyed to the inputs of a differential pair, a DC offset is induced which is proportional to the scalar product of the differential and common-mode component of the EMI signal that is injected in the input terminals. This is derived analytically in the corresponding paragraph. Several EMI resisting differential structures are reviewed and compared to each other in terms of EMI induced input offset, current consumption, matching constraints and input referred noise. Finally, a source-buffered differential pair generating a much smaller EMI induced offset voltage is introduced and described in detail. It is proven analytically that this structure (aside from generating a small EMI induced offset) has many advantages compared to the previously discussed structures: amongst others, it has a much more favorable noise behavior and it is more insensitive to process variations. In order to simulate and measure the EMI induced offset in a correct way, a measurement setup is introduced, which allows a correct EMI induced offset voltage measurement for a broad range of EMI frequencies. Previous observations were incorporated in the design of a test-IC. Measurements illustrate the superior electromagnetic immunity of the source-buffered differential pair opposed to the weak performance of the classic differential pair, even for substantial EMI levels.

Although the operational amplifier was one of the first analog integrated circuit blocks whose electromagnetic susceptibility has been studied, measured and reported, many misconceptions still exist concerning its immunity to EMI which is injected into the input terminals, as well as possible countermeasures and circuit improvements. This fact is accentuated by many publications which describe and focus on one particular immunity aspect of the opamp, without considering the other phenomena. This matter is complicated even more by proposed design solutions which were derived from solving one problem and which (sometimes unwittingly) turned out to be effective against the other phenomena. For instance, preventing slew rate asymmetries turned out to be effective against the nonlinearity of the input stage, as clarified in [Cor05]. This fact is of course benefic when improving designs [2], but unfortunately, complicates the exact comprehension of the EMI behavior of the global circuit. Even worse, small changes in the circuit topology or dimensioning may easily annihilate a superior EMI performance, because the dominant EMI effects are not recognized correctly, or as in many cases, are simply misunderstood. For this reason, all three phenomena will be described in detail.

## 2.1 Asymmetric slew rate

Slewing is an undesirable effect which takes place when large input signals are applied at the inputs of a given circuit (an operational amplifier in this case). While the small signal bandwidth of the opamp suggests a fast time domain response, the large signal speed is limited by the slew rate because the current which is maximally available to charge and to discharge the dominant capacitor in the circuit is not sufficient [Raz01]. As a consequence, during slewing, the relationship between the input and the output is heavily nonlinear and a lot of distortion is generated. As has been covered in Chap. 3, nonlinear distortion introduces rectification depending on the presence of even-order harmonics. At this point, a distinction must be made between positive and negative slew rate (respectively abbreviated as $SR+$ and $SR-$), meaning the maximum available current which is required to charge and respectively discharge the dominant

[2] Single solutions providing a global cure without consisting trade-offs can be categorized as an exceptional commodity in analog electronics.

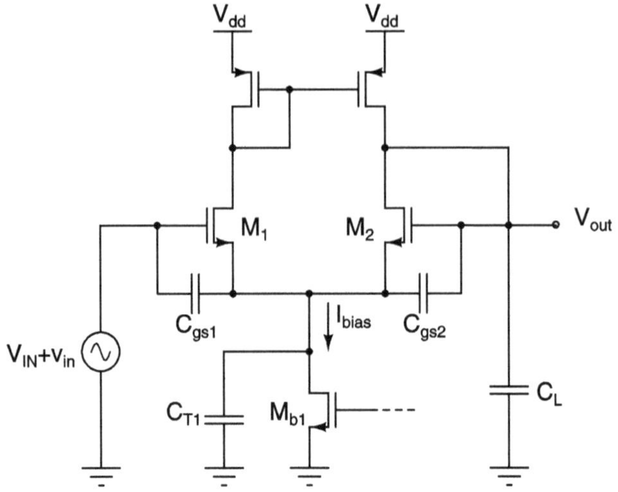


*Figure 5.1.* Basic one stage OTA connected as a voltage follower, with parasitic capacitors.

capacitor. Ideally, the positive and the negative slew rates are perfectly equal to each other, and the resulting nonlinearity is purely odd-ordered, since it is fully symmetrical.

Consider the basic one stage operational transconductance amplifier (OTA) connected in a voltage follower configuration as depicted in Fig. 5.1. The voltage follower configuration is often used in the literature when studying the EMI induced asymmetric slew rate effect, because this way, the voltage swings on the input differential pair are the highest, as reported in [Mas96]. Ideally, the positive and negative slew rates of this OTA are equal to each other, and defined by the bias current ($I_{bias}$) and the load capacitor ($C_L$):

$$SR = SR+ = SR- = \frac{I_{bias}}{C_L} \tag{5.1}$$

Unfortunately, both the positive and the negative slew rates are seldom perfectly equal to each other. Detailed analyses for various opamps have been developed describing analytically the divergences between positive and negative slew rate, e.g. in two stage Miller opamps as in [Yav05]. The effect of slew rate asymmetries on the circuit's susceptibility to EMI has amongst others been analyzed in [Gra97] and [Mas99], where it has been illustrated that the inequality between $SR+$ and $SR-$ mainly depend on three effects:

- Charge modulation across transistor $M_{b1}$ providing bias current $I_{bias}$. Assume that a positive voltage step is applied at the input of the OTA connected in a voltage follower configuration as depicted in Fig. 5.1. Doing so, the voltage at the source of $M_1$ increases, forcing the bias current to a higher value owing to the channel length modulation of $M_{b1}$. This current is slightly higher than its quiescent value, and it is used to charge capacitor $C_L$. Conversely, a negative voltage step applied at the gate of $M_1$ decreases the output voltage $V_{out}$, and at the same time the gate of $M_2$. This means that owing to the channel length modulation of $M_{b1}$, the bias current delivered by this transistor decreases [Gra97].

- Asymmetries in the circuit topology, causing the dominant capacitor to charge faster than it is able to discharge, and vice versa [Gra97].
- Parasitic capacitances, and in particular, the parasitic gate-source capacitances associated to the input differential pair ($C_{gs1}$ and $C_{gs2}$) as well as the parasitic capacitance $C_{T1}$ coupling the sources of $M_1$ and $M_2$ to ground [Mas99].

As a consequence, the positive and negative slew rates are not equal to each other: variations of 10% up to 20% have been reported in [Gra97]. Asymmetric slew rates can generate DC shift because asymmetries cause even-order nonlinear distortions. This is basically the same phenomenon that has been described with the source follower example in Chap. 3, in which the voltage across the output capacitor is pumped to a higher or lower value, because the charge and the discharge paths are not equal to each other. This effect is illustrated in Fig. 5.2, where the pumping effect owing to an unequal positive and negative slew rate is demonstrated for the basic one stage OTA represented in Fig. 5.1. Following guidelines have been developed in reference works in order to obtain a high correspondence between $SR+$ and $SR-$:

- Minimizing the parasitic capacitances of the input differential stage, in particular the parasitic capacitance across the tail current source transistor ($C_{T1}$) [Mas99].
- Minimizing the channel length modulation effect of tail current transistor $M_{b1}$ by increasing its channel length [Gra97].
- Observing a high topological symmetry by using fully differential topologies and mirrored signal paths. In [Ric01], a fully differential folded cascode amplifier structure has been connected to a cross-coupled and highly symmetrical class AB buffer, leading to a highly symmetrical slew rate. In this same design, extra internal capacitors between the gate and the source of $M_1$ and $M_2$ have been added, which improves the linearity of the input differential pair, as is derived in the following paragraphs. Although focused on slew rate symmetrization, this solution turned out to be effective against the nonlinearity of the input differential stage as well, as stated in [Cor05].

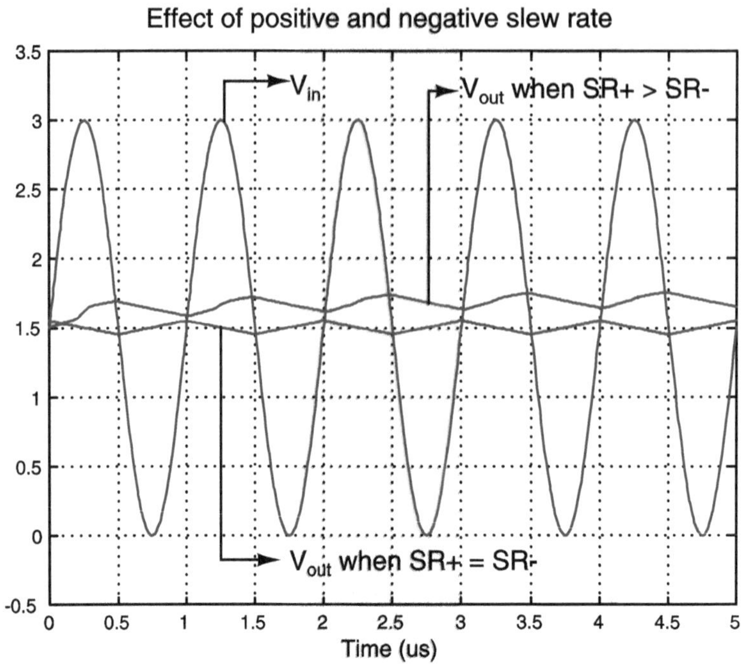


*Figure 5.2.* Effect of unequal positive and negative slew rates in a basic one stage OTA.

Finally, slew rate asymmetry plays a major role for low to medium EMI frequencies, situated around the unity gain frequency of the opamp. This has been certified with measurements and calculations in [Gra97] and [Mas96]. At very high EMI frequencies, no slew rate induced DC shift occurs, since the input signal is filtered by the parasitic capacitances of the input transistors.

## 2.2 Strong nonlinear behavior of the input differential pair

Parasitic capacitances become more relevant at high frequencies, in particular at frequencies lying well above the bandwidth of the amplifier. It has been observed in [Gra97] that owing to the former, the input stages of slew rate optimized opamp designs still exhibit asymmetric behavior during transients when they are driven by high frequency, high amplitude EMI signals. This is illustrated as follows. Consider the elementary one stage OTA connected as a voltage follower, which was depicted in Fig. 5.1. At very high EMI frequencies, above the amplifier's unity gain frequency, the parasitic capacitances of the differential input pair are predominant. In addition, since no amplification takes place at those high EMI frequencies, the output of the opamp is a quasi DC signal. In addition, the AC drain currents of $M_1$ and $M_2$ are negligible since most of the AC current is flowing through the parasitic capacitors. Assuming that the gate-source capacitances of $M_1$ and $M_2$ are equal to each other ($C_{gs1} = C_{gs2} = C_{gs}$), the gate-source voltages of $M_1$ and $M_2$ are established as follows [Gra97]:

$$\begin{cases} v_{gs1}(t) = v_{in}(t)\frac{C_{gs}+C_{T1}}{2C_{gs}+C_{T1}} \\ v_{gs2}(t) = -v_{in}(t)\frac{C_{gs}}{2C_{gs}+C_{T1}} \end{cases} \tag{5.2}$$

Previous relationship shows that the magnitude of the AC current through $M_1$ is larger than the magnitude of the AC current through $M_2$, since $|v_{gs1}(t)| > |v_{gs2}(t)|$. This means that a large sinusoidal input voltage $v_{in}(t)$ draws $M_1$ in cut-off for a longer time than $M_2$ [Mas99]. Clearly, based on previous relationship, the distortion in the drain current of $M_1$ is larger than the one in $M_2$. The DC shift on the output equivalently converges towards a value which is characterized by the DC value of the gate source voltages, which are themselves a function of the parasitic gate-source capacitances of $M_1$ and $M_2$, as well as of the shunt capacitor across the tail current source transistor $M_{b1}$, as specified in (5.2). The conclusion is that the magnitudes of both gate-source voltages ($|v_{gs1}|$ and $|v_{gs2}|$) must be made equal to each other. This is accomplished by minimizing the parasitic capacitor $C_{T1}$, and by increasing the gate-source capacitances of the input differential pair, as suggested in [Gra97]. The former criterion overlaps with the conclusions observed in the previous paragraph when describing ways to minimize the asymmetric slew rate. Coincidentally, adding extra gate-source capacitances while minimizing $C_{T1}$ improves the global linearity of the input differential pair as well, as is illustrated analytically in the next paragraph.

## 2.3 Weak nonlinear behavior of the input differential pair

### 2.3.1 EMI induced offset in a classic differential pair

When a classic differential pair is driven simultaneously by a common-mode and a differential-mode EMI disturbance (Fig. 5.3), an output offset current is generated owing to the finite impedance which is present between the sources of $M_1$-$M_2$, and ground. This impedance comprises the output resistance of current source $M_{b1}$ (which will be neglected), and the parasitic capacitance between the drain of $M_{b1}$ and ground [Fio02a, Fio06]. The latter is comprised of (refer to Fig. 5.4):

- The parasitic drain-bulk capacitance of transistor $M_{b1}$ ($C_{db}$).
- The parasitic junction capacitance between the bulk and the isolating well of input transistors $M_1$ and $M_2$ ($C_j$). This capacitor can be eliminated by connecting the bulk of $M_1$ and $M_2$ to the substrate. However, this increases the threshold voltage $V_t$, which is usually not desirable: in addition, substrate noise affects the good working of MOS transistors through the body

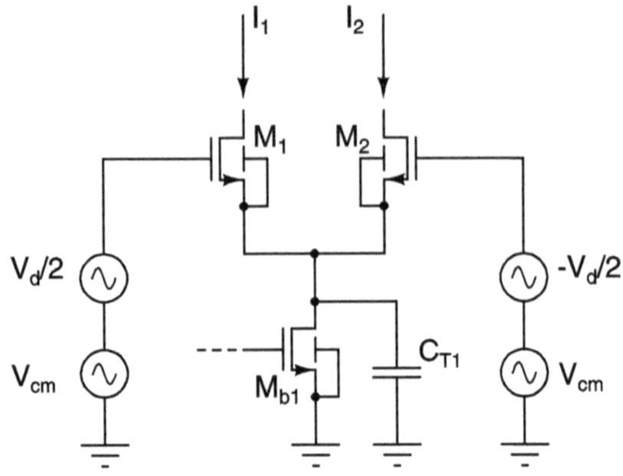


*Figure 5.3.* Classic differential pair, disturbed by a common-mode and a differential-mode EMI signal of the same frequency. Because of $C_{T1}$, $\overline{I_{d1}}$ is no longer equal to $\overline{I_{d2}}$, generating a net offset current.

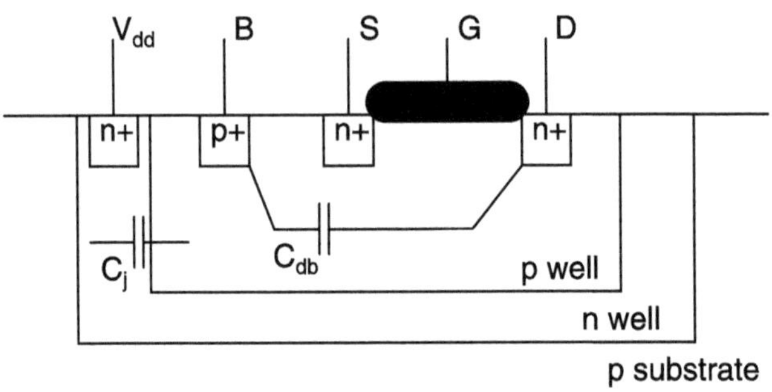


*Figure 5.4.* Cross section of a NMOS transistor, displaying its parasitic $C_{db}$ and $C_j$ capacitances.

effect, and varies the threshold voltage as well as the bulk transconductance ($g_{mb}$) [Zin00].

The sum of both capacitors is represented as $C_{T1}$, which is connected between the sources of $M_1$-$M_2$ and ground. This capacitor is responsible for the EMI induced DC shift. This is now derived mathematically. Using the expression for a MOS transistor in saturation [Raz01], and expressing the total gate-source voltage $V_{gsi}$ as the sum of a DC ($V_{GSi}$) and an AC component ($v_{gsi}$), the output offset current ($I_{OS}$) is expressed as:

$$\begin{aligned} I_{OS} &= \overline{I_1} - \overline{I_2} \\ &= \beta \cdot \overline{(V_{gs1} - V_t)^2} - \beta \cdot \overline{(V_{gs2} - V_t)^2} \\ &= \beta \cdot (\overline{v_{gs1}^2} - \overline{v_{gs2}^2}) \end{aligned} \quad (5.3)$$

where $\beta = \frac{\mu C_{ox}}{2}\frac{W}{L}$. The Parseval identity for Fourier integrals states that the total energy contained in a transient waveform summed across all of time $t$ is equal to the total energy of the waveform's Fourier transform summed across all of its frequency components [Arf95, Spi74, Spi94]. This results in the following identity:

$$\overline{v_{gsi}^2(t)} = \lim_{T\to\infty}\frac{1}{T}\int_{\frac{-T}{2}}^{\frac{T}{2}} v_{gsi}^2(t)\cdot dt = \int_{-\infty}^{\infty}\left|V_{gsi}(j\omega)\right|^2\cdot d\omega \tag{5.4}$$

Consequently, the output offset current can be expressed in the frequency domain as follows:

$$I_{OS} = \beta\cdot\int_{-\infty}^{\infty}\left|V_{gs1}(j\omega)\right|^2\cdot d\omega - \beta\cdot\int_{-\infty}^{\infty}\left|V_{gs2}(j\omega)\right|^2\cdot d\omega \tag{5.5}$$

Expressing $V_{gsi}(j\omega)$ in terms of the common-mode ($V_{cm}(j\omega)$) and the differential-mode ($V_{dm}(j\omega)$) input voltage yields:

$$\begin{cases} V_{gs1}(j\omega) = H_{cm}(j\omega)\cdot V_{cm}(j\omega) + \frac{V_{dm}(j\omega)}{2} \\ V_{gs2}(j\omega) = H_{cm}(j\omega)\cdot V_{cm}(j\omega) - \frac{V_{dm}(j\omega)}{2} \end{cases} \tag{5.6}$$

where $H_{cm}(j\omega)$ is the transfer function for common-mode signals. Provided that $g_{m1} = g_{m2}$, $C_{gs1} = C_{gs2}$, $H_{cm}(j\omega)$ can be expressed as:

$$H_{cm}(j\omega) = \frac{j\omega\cdot C_{T1}}{2\cdot g_{m1} + j\omega\cdot(C_{T1} + 2C_{gs1})} \tag{5.7}$$

Rewriting (5.5) using (5.6) and (5.7) in terms of the input offset voltage ($V_{OS} = I_{OS}/g_{m1}$) yields:

$$\begin{aligned} V_{OS} &= \frac{I_{OS}}{g_{m1}} \\ &= \frac{\beta}{g_{m1}}\cdot\int_{-\infty}^{\infty}\left|V_{gs1}(j\omega)\right|^2\cdot d\omega - \frac{\beta}{g_{m1}}\cdot\int_{-\infty}^{\infty}\left|V_{gs2}(j\omega)\right|^2\cdot d\omega \\ &= \frac{2\cdot\beta}{g_{m1}}\cdot\int_{-\infty}^{\infty}\mathrm{Re}\{H_{cm}(j\omega)\cdot V_{cm}(j\omega)\cdot V_{dm}(j\omega)\}\cdot d\omega \\ &= \frac{1}{V_{GS1} - V_t}\int_{-\infty}^{\infty}\left|H_{cm}(j\omega)\cdot V_{cm}(j\omega)\cdot V_{dm}(j\omega)\right|\cdot\cos\phi\cdot d\omega \end{aligned} \tag{5.8}$$

where:

$$\phi = \mathrm{atan}\left(\frac{\mathrm{Im}\{H_{cm}(j\omega)\cdot V_{cm}(j\omega)\cdot V_{dm}(j\omega)\}}{\mathrm{Re}\{H_{cm}(j\omega)\cdot V_{cm}(j\omega)\cdot V_{dm}(j\omega)\}}\right) \tag{5.9}$$

Equation (5.8) lists the main contributing elements to the EMI induced input offset voltage of the classic differential pair. Three observations can be made:

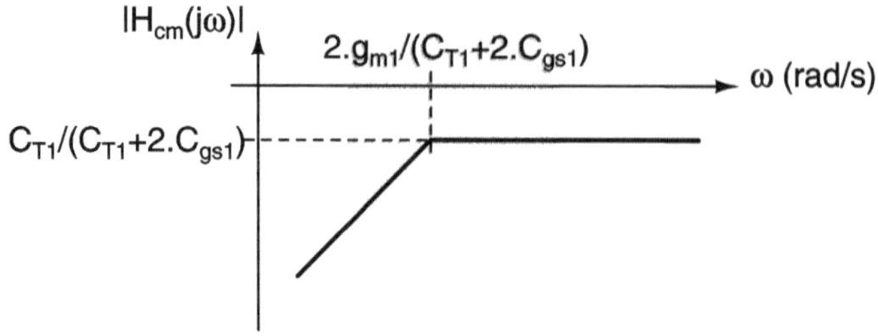


*Figure 5.5.* Simplified Bode plot of $|H_{cm}(j\omega)|$ in a classic differential pair.

- The input offset voltage is proportional to the multiplication of the magnitudes of the common-mode and differential-mode EMI components, as well as to the phase between both.

- The larger the overdrive voltage ($V_{GS1} - V_t$), the smaller the input offset voltage: this is a controversial requirement, since the overdrive voltage of an input differential pair must preferably be as small as possible in order to reduce the offset which is caused by mismatch [Raz01]. It is therefore difficult to look for a convenient optimum using $V_{GS1} - V_t$ as a design parameter. Additionally, the EMI induced offset can typically lie many orders of magnitude above the offset caused by mismatch, meaning that a sound trade-off or optimization does not yield practical results.

- In order to decrease the offset current, $|H_{cm}(j\omega)|$ must be reduced as much as possible: this can be realized by increasing the gate-source capacitances $C_{gs1}$ and $C_{gs2}$, and by decreasing $C_{T1}$ (refer to (5.7)). This corresponds to what has been derived previously, in Sects. 2.1 and 2.2: at the same time, it proves why previous solutions focusing on eliminating the asymmetric slew rate and the strong nonlinear behavior owing to parasitic capacitances by connecting extra gate-source capacitances, turned out to be effective against this small signal nonlinear behavior as well. It is, however, not possible to make $H_{cm}(j\omega)$ zero owing to parasitic effects. A simplified Bode plot of $|H_{cm}(j\omega)|$ is shown in Fig. 5.5.

*Example*: Consider the special case where the differential-mode as well as the common-mode EMI components are two sinusoids:

$$\begin{aligned} V_{dm}(t) &= \hat{V}_{dm} \cdot \cos(\omega_o t) \\ V_{cm}(t) &= \hat{V}_{cm} \cdot \cos(\omega_o t + \theta) \end{aligned} \tag{5.10}$$

The Fourier transformation of both time domain signals yields:

$$V_{dm}(j\omega) = \frac{\hat{V}_{dm}}{2} \cdot (\delta(\omega - \omega_o) + \delta(\omega + \omega_o))$$
$$V_{cm}(j\omega) = \frac{\hat{V}_{cm}}{2} \cdot \left(\delta(\omega - \omega_o) + \delta(\omega + \omega_o)\right) \tag{5.11}$$

where $\delta$ represents the Dirac's delta function [Arf95]. In this particular case, (5.8) is rewritten as:

$$V_{OS} = \frac{\hat{V}_{dm} \cdot \hat{V}_{cm}}{4 \cdot (V_{GS1} - V_t)} \int_{-\infty}^{\infty} \text{Re}\{H_{cm}(j\omega) \cdot (\delta^2(\omega - \omega_o) + \delta^2(\omega + \omega_o))\} d\omega$$

$$= \frac{\hat{V}_{dm} \cdot \hat{V}_{cm} \cdot |H_{cm}(j\omega)| \cdot \cos(\phi_{H_{cm}} + \theta)}{2 \cdot (V_{GS1} - V_t)} \tag{5.12}$$

where:

$$\phi_{H_{cm}} = \text{atan}\left(\frac{\text{Im}\{H_{cm}\}}{\text{Re}\{H_{cm}\}}\right) \tag{5.13}$$

Observing (5.12) and (5.13), the highest offset voltage occurs for EMI frequencies which lie above the pole frequency of $H_{cm}(j\omega)$. For these frequencies, $\phi_{H_{cm}}$ is zero. In the worst case situation being that the common-mode and the differential-mode EMI components are in phase ($\theta = 0°$) and situated at high EMI frequencies (well above the pole frequency of $H_{cm}(j\omega)$), the maximal worst case input offset voltage is equal to:

$$V_{OS\_max} = \frac{\hat{V}_{dm} \cdot \hat{V}_{cm}}{2 \cdot (V_{GS1} - V_t)} \cdot \left|\frac{C_{T1}}{C_{T1} + 2 \cdot C_{gs1}}\right| \tag{5.14}$$

Finally, proposed solutions will have a significant impact on the equivalent noise voltage of a classic differential pair, as will be illustrated in the subsequent paragraphs. In case the input transistors are equal in size ($g_{m1} = g_{m2}$), the equivalent thermal input noise per unit bandwidth of a classic differential pair is expressed as follows [3][Raz01]:

$$\overline{V_{n,in}^2} = 8kT \cdot \left(\frac{2}{3 \cdot g_{m1}}\right) \tag{5.15}$$

[3] To be perfectly accurate, the equivalent thermal input noise per unit bandwidth is equal to $\overline{V_{n,in}^2} = 8kT \cdot (\frac{\gamma}{g_{m1}})$, where $\gamma$ is equal to 2/3 for long channel devices [Raz01].

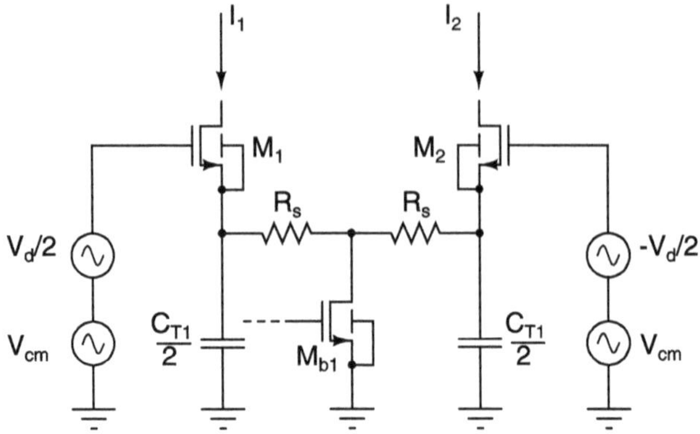


*Figure 5.6.* Classic differential pair with resistive source degeneration, disturbed by a common-mode and a differential-mode EMI signal of the same frequency.

### 2.3.2 Classic differential pair using source degeneration

Since distortion phenomena are responsible for the EMI induced DC shift in classic differential pairs, a possible solution is to linearize the differential pair by adding source degeneration resistors as proposed in [Cor05]. Refer to Fig. 5.6. Observe that the drain-bulk capacitor of $M_{b1}$ has been disregarded, compared to the much larger bulk-isolation well capacitance of $M_1$-$M_2$. Similarly to (5.6), the gate-source voltage $V_{gsi}(j\omega)$ is expressed in terms of the common-mode ($V_{cm}(j\omega)$) and the differential-mode ($V_{dm}(j\omega)$) input voltage:

$$\begin{cases} V_{gs1}(j\omega) = H_{cm}(j\omega) \cdot V_{cm}(j\omega) + H_{dm}(j\omega) \cdot \frac{V_{dm}(j\omega)}{2} \\ V_{gs2}(j\omega) = H_{cm}(j\omega) \cdot V_{cm}(j\omega) - H_{dm}(j\omega) \cdot \frac{V_{dm}(j\omega)}{2} \end{cases} \tag{5.16}$$

where $H_{cm}(j\omega)$ and $H_{dm}(j\omega)$ represent the common-mode and differential-mode transfer function. $H_{cm}(j\omega)$ is identical to (5.7), since the common-mode transfer function is similar to the one calculated in the previous paragraph. The differential-mode transfer function $H_{dm}(j\omega)$ is expressed as:

$$H_{dm}(j\omega) = \frac{2 + C_{T1} \cdot R_s \cdot j\omega}{(2 + 2 \cdot R_s \cdot g_{m1}) + (2 \cdot C_{gs1} \cdot R_s + C_{T1} \cdot R_s) \cdot j\omega} \tag{5.17}$$

The input offset is now expressed as follows:

$$V_{OS} = \frac{1}{V_{GS1} - V_t} \int_{-\infty}^{\infty} \big| H_{cm}(j\omega) \cdot V_{cm}(j\omega) \cdot H_{dm}(j\omega) \cdot V_{dm}(j\omega) \big| \cdot \cos\phi \cdot d\omega \tag{5.18}$$

where:

$$\phi = \operatorname{atan}\left( \frac{\operatorname{Im}\{H_{cm}(j\omega) \cdot V_{cm}(j\omega) \cdot H_{dm}(j\omega) \cdot V_{dm}(j\omega)\}}{\operatorname{Re}\{H_{cm}(j\omega) \cdot V_{cm}(j\omega) \cdot H_{dm}(j\omega) \cdot V_{dm}(j\omega)\}} \right)$$

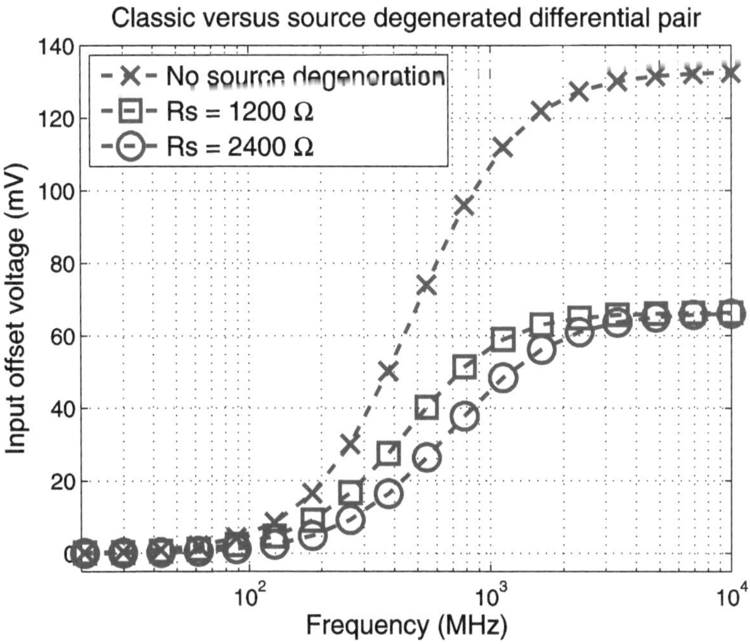


*Figure 5.7.* Input referred offset voltage plotted against frequency, for different values of degeneration resistor $R_s$. Observe that the offset voltage decreases for increasing $R_s$.

Clearly, since $|H_{dm}(j\omega)| < 1$, the EMI induced input offset voltage is smaller than the one expressed in (5.8), depending on the value of degeneration resistors $R_s$. This offset reduction does not come for free however. First of all, the total transconductance decreases. In addition, the equivalent input noise is increased because of the presence of the degeneration resistors in the signal path:

$$\overline{V_{n,in}^{\ 2}} = 8kT \cdot \left( \frac{2}{3 \cdot g_{m1}} + R_s \right) \tag{5.19}$$

*Example*: Consider the following case: $g_{m1} = g_{m2} = 710$ µS, $C_{gs1} = C_{gs2} = 117$ fF, $C_{T1} = 234$ fF. The resulting input offset voltage has been plotted for different values of $R_s$ in Fig. 5.7. The injected EMI signal level equals $V_d = 200$ mV (RMS) and $V_{cm} = 100$ mV (RMS).

### 2.3.3 Cross-coupled differential pair

In [Fio01], a cross-coupled differential pair has been presented which compensates for the EMI induced offset in a differential pair by summing two complementary average drain currents, hereby zeroing their net offset contribution.

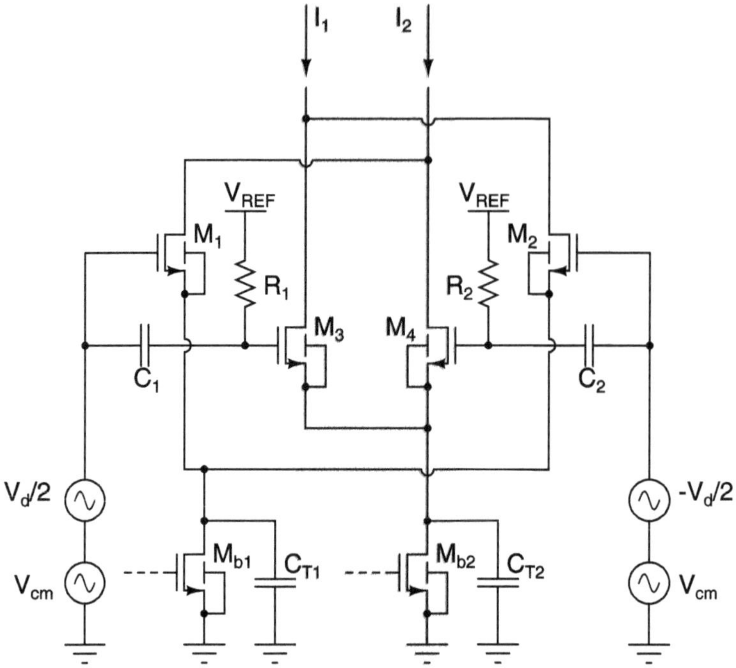


*Figure 5.8.* Cross-coupled differential pair, disturbed by a common-mode and a differential-mode EMI signal of the same frequency.

This circuit is depicted in Fig. 5.8: it is composed of two differential pairs ($M_1$-$M_2$ and $M_3$-$M_4$), which are cross-coupled at the drains. The DC voltage at the gate of $M_3$-$M_4$ is fixed by a reference voltage source ($V_{REF}$), situated ideally just in the middle of the common-mode voltage range of $M_1$-$M_2$. Two matched $R$-$C$ high-pass filters couple the input signal to the gates of $M_3$-$M_4$. Transistor pairs $M_1$-$M_2$ and $M_3$-$M_4$, as well as resistors $R_1$-$R_2$ and capacitors $C_1$-$C_2$ are ideally matched to each other. The main idea behind this solution is that the extra differential pair $M_3$-$M_4$ becomes active only in the pass-band of the $R$-$C$ high-pass filter: the cut-off frequency of this filter must therefore lie above the wanted frequency band. Subtracting the offset current generated by $M_3$-$M_4$ from the original offset current (caused by $M_1$-$M_2$) by way of cross-coupling creates a circuit which is in principle free of DC shift. This is now derived mathematically. Provided that the four transistors in this circuit are ideally matched to each other ($g_{m1} = g_{m2} = g_{m3} = g_{m4}$), and provided that they are biased identically by perfectly equal current sources ($I_{b1} = I_{b2}$ and $C_{T1} = C_{T2}$), the resulting theoretical input offset voltage can be expressed as follows:

$$\begin{aligned}V_{OS} &= \frac{I_{OS\_M1M2} - I_{OS\_M3M4}}{g_{m1}} \\ &= \frac{1}{V_{GS1} - V_t} \int_{-\infty}^{\infty} |H_{cm}(j\omega) \cdot V_{cm}(j\omega) \cdot V_{dm}(j\omega)| \\ &\quad \cdot \left(1 - \left|\frac{j\omega \cdot R_1 \cdot C_1}{1 + j\omega \cdot R_1 \cdot C_1}\right|^2\right) \cdot \cos\phi \cdot d\omega \end{aligned} \tag{5.20}$$

Previous expression can be simplified to:

$$\begin{aligned}V_{OS} &= \frac{1}{V_{GS1} - V_t} \int_{-\infty}^{\infty} |H_{cm}(j\omega) \cdot V_{cm}(j\omega) \cdot V_{dm}(j\omega)| \\ &\quad \cdot \left|\frac{1}{1 + j\omega \cdot R_1 \cdot C_1}\right|^2 \cdot \cos\phi \cdot d\omega \end{aligned} \tag{5.21}$$

Clearly, for high EMI frequencies, $V_{OS}$ is close to zero, meaning that there is no net EMI induced offset.

There are, however, many drawbacks related to this structure. First of all, it has an increased susceptibility to noise owing to the extra transistor pair and owing to the resistors used in the $R$-$C$ high-pass filter. The total equivalent input noise is now equal to:

$$\overline{V_{n,in}^{\ 2}} = 8kT \cdot \left(\frac{2}{3 \cdot g_{m1}} + \frac{2}{3 \cdot g_{m3}} + R_1\right) \tag{5.22}$$

In case the transistors are equal in size and biased identically, $g_{m1} = g_{m3}$ and this yields:

$$\overline{V_{n,in}^{\ 2}} = 8kT \cdot \left(\frac{4}{3 \cdot g_{m1}} + R_1\right) \tag{5.23}$$

Extra disadvantages associated to the cross-coupled differential pair structure include an increased power dissipation and a larger area. More importantly, this differential pair structure is very liable to mismatch because it is strongly dependent on a perfect subtraction of generated EMI induced offsets. The sensitivity to mismatch of this structure has been illustrated in [Wal07] using Monte Carlo simulations. Moreover, this structure requires an extra reference voltage $V_{REF}$, equaling the input common-mode voltage of $M_1$-$M_2$: this complicates the biasing of this circuit. Additionally, if $V_{REF}$ is not equal to the DC bias voltage of $M_1$-$M_2$, the offset compensation drastically worsens. Finally, a variation to this circuit, relying on the subtraction of offset currents has been published in [Fio02b]. In the latter reference, the cross-coupling of NMOS differential pairs has been replaced with a PMOS and a NMOS differential pair cancelling each other's offset current. However, this structure depends heavily

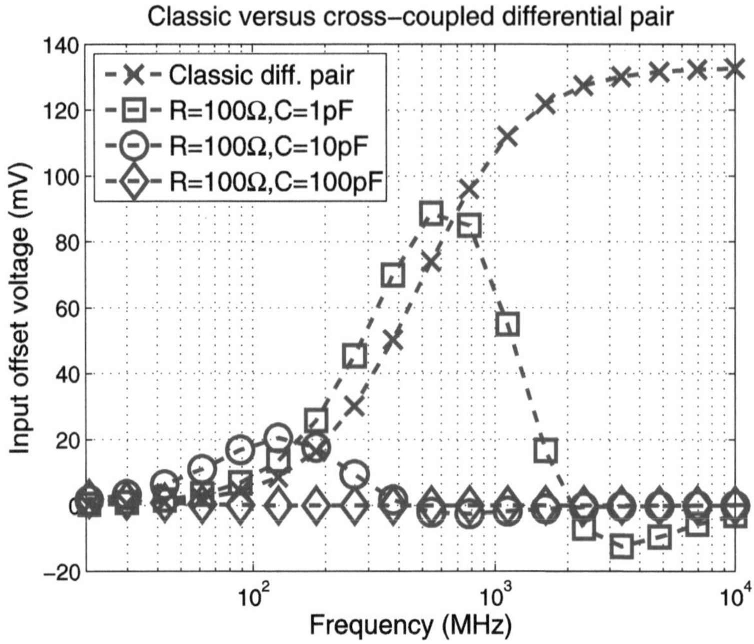


*Figure 5.9.* Input referred offset voltages of the cross-coupled differential pair and the classic differential pair plotted against frequency, for different values of $R_1$-$C_1$. Observe that the offset decreases for increasing values of $R_1$-$C_1$.

on the matching between PMOS and NMOS transistors, which is very difficult to achieve.

*Example*: Consider the following case: $g_{m1} = g_{m2} = g_{m3} = g_{m4} = 710$ µS, $C_{gs1} = C_{gs2} = C_{gs3} = C_{gs4} = 117$ fF, $C_{T1} = C_{T2} = 234$ fF. The resulting input offset voltage has been plotted for different values of $R_1$-$C_1$ in Fig. 5.9. The injected EMI signal level equals $V_d = 200$ mV RMS and $V_{cm} = 100$ mV RMS as previously. Observe that for high values of $R_1$-$C_1$, the input referred offset is close to zero.

### 2.3.4 Differential pair with low-pass $R$-$C$ filter

In [Wal07], a more effective approach is proposed. Refer to Fig. 5.10. In this circuit, a $R$-$C$ low-pass filter is placed in front of the differential pair, effectively rejecting any out-of-band common-mode and differential-mode EMI disturbances which are superposed on the input signal. This solution seems trivial, however, in case this $R$-$C$ low-pass filter is dimensioned with an identical cut-off frequency as the $R$-$C$ high-pass filter which is used to drive the cross-coupled differential pair described in the previous section, both the cross-coupled differential pair and the solution using a low-pass filter in front of the

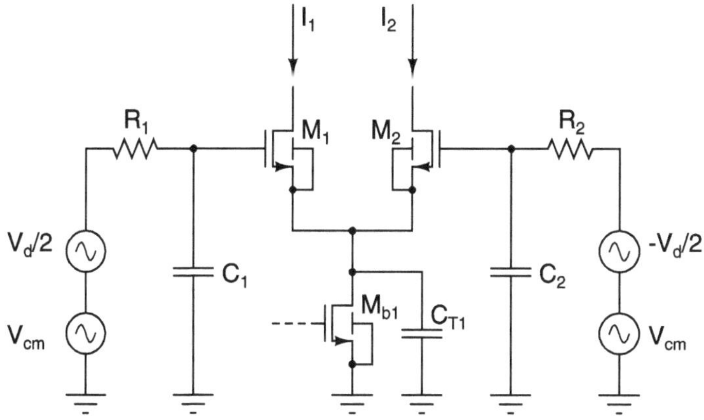


*Figure 5.10.* Differential pair with $R$-$C$ low-pass filter, disturbed by a common-mode and a differential-mode EMI signal of the same frequency.

differential pair generate the same theoretical DC shift. This is illustrated as follows. The resulting input offset voltage of the $R$-$C$ high-pass filtered differential pair (Fig. 5.10) can indeed be expressed as:

$$V_{OS} = \frac{I_{OS}}{g_{m1}} = \frac{1}{V_{GS1} - V_t} \int_{-\infty}^{\infty} |H_{cm}(j\omega) \cdot V_{cm}(j\omega) \cdot V_{dm}(j\omega)| \cdot \left| \frac{1}{1 + j\omega \cdot R_1 \cdot C_1} \right|^2 \cdot \cos\phi \cdot d\omega \tag{5.24}$$

A quick inspection reveals that both offsets expressed in (5.21) and (5.24) are similar. Moreover, simulations indicate that the differential pair with low-pass filtering generates less DC shift: indeed, since the common-mode and differential input voltages are attenuated before reaching the gates of the input transistor pair, the voltage swings at the gates and sources of the input transistors are smaller [Wal07]. This results in a more accurate small signal approximation, and additionally, less influence from the asymmetric slew rate phenomenon (Sect. 2.1) and large nonlinearities (Sect. 2.2). Further keeping in mind the previously cited disadvantages, like an increased power consumption of the cross-coupled differential pair, as well as a strong dependence on matching and the necessity of an extra bias voltage, preference is given to the low-pass input filter differential pair structure. In addition, the noise behavior is improved. The total equivalent input noise is equal to:

$$\overline{V_{n,in}{}^2} = 8kT \cdot \left( \frac{2}{3 \cdot g_{m1}} + R_1 \right) \tag{5.25}$$

The MOS transistor channel noise contribution to the total noise is evidently halved compared to (5.23), since two input transistors are used instead of four.

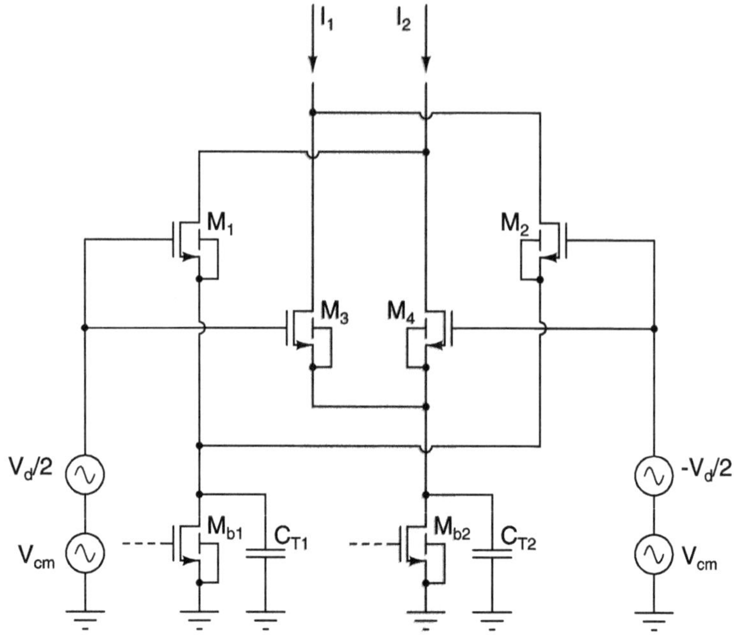


*Figure 5.11.* Improved cross-coupled differential pair, disturbed by a common-mode and a differential-mode EMI signal of the same frequency.

Leaving out the capacitors $C_1$ and $C_2$, results in the solution which was proposed in [Cor05]. However, this results in even larger resistors $R_1$ and $R_2$ in order to achieve the same cut-off frequency, which therefore introduce even more noise as can be observed in (5.25). Clearly, the main noise contribution stems from the thermal noise associated to resistor $R_1$. The next evident step is to get rid of this resistor.

### 2.3.5 Improved cross-coupled differential pair

An improvement of the original cross-coupled differential pair which was described in Sect. 2.3.3 has been reported in [Fio06]. The principle behind this circuit stems on using a different transistor sizing, instead of relying on a high-pass filter. Refer to Fig. 5.11. Using an analogous approach, the offset output current is calculated as follows:

$$\begin{aligned} I_{OS} = \beta_1 \cdot & \int_{-\infty}^{\infty} |H_{cm1}(j\omega) \cdot V_{cm}(j\omega) \cdot V_{dm}(j\omega)| \cdot \cos\phi \cdot d\omega \\ & - \beta_3 \cdot \int_{-\infty}^{\infty} |H_{cm3}(j\omega) \cdot V_{cm}(j\omega) \cdot V_{dm}(j\omega)| \cdot \cos\phi \cdot d\omega \qquad (5.26) \end{aligned}$$

where:

$$\begin{cases} \beta_i = \frac{\mu C_{ox}}{2} \frac{W_i}{L_i} \\ H_{cm1}(j\omega) = \frac{j\omega \cdot C_{T1}}{2 \cdot g_{m1} + j\omega \cdot (C_{T1} + 2C_{gs1})} \\ H_{cm3}(j\omega) = \frac{j\omega \cdot C_{T2}}{2 \cdot g_{m3} + j\omega \cdot (C_{T2} + 2C_{gs3})} \end{cases} \quad (5.27)$$

Theoretically, this means that the output offset current is zero, provided that:

$$\beta_1 \cdot H_{cm1}(j\omega) = \beta_3 \cdot H_{cm3}(j\omega) \quad (5.28)$$

Obviously, previous condition is satisfied in case all four transistors and biasing currents are identical, but this choice results in a trivial solution of a zero AC output current for any inputs. For this reason, the condition specified in (5.28) must be satisfied for $g_{m1} \neq g_{m3}$. Observing previous comments, the two following equations are derived [Fio06]:

$$\begin{cases} \frac{\beta_1 \cdot C_{T1}}{C_{T1} + 2 \cdot C_{gs1}} = \frac{\beta_2 \cdot C_{T2}}{C_{T2} + 2 \cdot C_{gs3}} \\ \frac{g_{m1}}{C_{T1} + 2 \cdot C_{gs1}} = \frac{g_{m3}}{C_{T2} + 2 \cdot C_{gs3}} \end{cases} \quad (5.29)$$

Using both conditions, differential pair $M_3$-$M_4$ can be designed in order to compensate the offset current which is generated by $M_1$-$M_2$ owing to the EMI distortion. Since the transconductance of the initial differential stage ($M_1$-$M_2$) is reduced by the secondary differential pair ($M_3$-$M_4$), it is important to keep $g_{m3}$ at a low value, by reducing the bias current flowing through $M_3$-$M_4$. In theory, if all the circuit parameters are exactly known and dimensioned as prescribed in (5.29), this circuit generates no DC shift. However, the influence of mismatch is very pronounced, and the smallest deviation from the calculated theoretical values yields a significant input offset voltage. In addition, the extra differential pair increases the total noise of the input stage, since $g_{m3}$ is much smaller than $g_{m1}$. The equivalent input noise voltage of the cross-coupled differential pair is equal to:

$$\overline{V_{n,in}}^2 = 8kT \cdot \left( \frac{2}{3 \cdot g_{m1}} + \frac{2}{3 \cdot g_{m3}} \right) \quad (5.30)$$

The advantage compared to previous differential pair structures is that there is no resistor in the signal path, meaning that the input noise voltage is reduced. However, since $g_{m3}$ is much smaller than $g_{m1}$, the total referred input noise can still be considerable.

### 2.3.6 Source-buffered differential pair

In [Red06c], a source-buffered differential pair is presented: refer to Fig. 5.12. The nominal differential pair ($M_1$-$M_2$) is back-biased by an auxiliary differential pair ($M_3$-$M_4$). The latter bootstraps the bulk-source voltage of $M_1$-$M_2$ and keeps the average drain current of $M_1$ and $M_2$ at a constant value by forcing

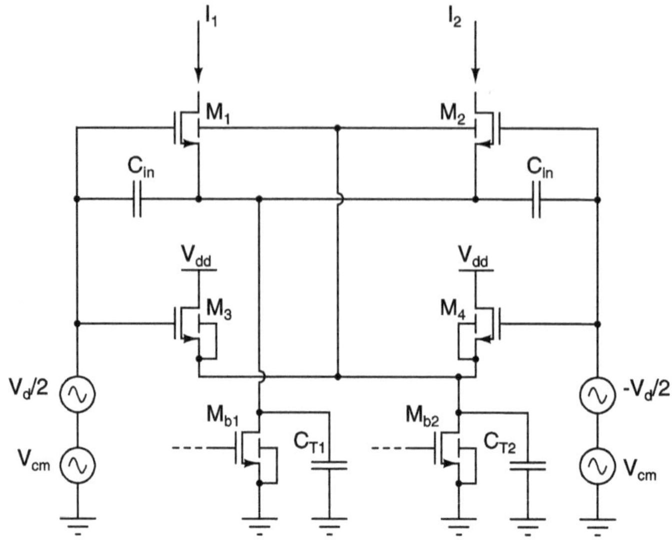


*Figure 5.12.* Source-buffered differential pair, disturbed by a common-mode and a differential-mode EMI signal of the same frequency.

the common-mode transfer function $H_{cm}(j\omega)$ to zero. First observe that the output offset current is expressed as a function of the gate-source ($V_{gs}$) and the source-bulk voltage ($V_{sb}$). Using the expression for a MOS transistor in saturation, and expressing $V_{gsi}$, $V_{sbi}$ as the sum of a DC ($V_{GSi}$, $V_{SBi}$) and an AC component ($v_{gsi}$, $v_{sbi}$), the output offset current ($I_{OS}$) is expressed as:

$$I_{OS} = \beta \cdot \left( \overline{(V_{gs1} - V_t)^2} - \overline{(V_{gs2} - V_t)^2} \right) \tag{5.31}$$

where $\beta = \frac{\mu C_{ox}}{2} \frac{W}{L}$. Since the bulk of $M_1$ and $M_2$ is not connected to the source, the threshold voltage $V_t$ is no longer constant and the body effect must be included [Raz01]. This yields:

$$\begin{aligned} I_{OS} = \beta \cdot \Big( & \overline{(V_{gs1} - V_{t0} - \gamma\sqrt{2\Phi_F + V_{sb1}} + \gamma\sqrt{2\Phi_F})^2} \\ & - \overline{(V_{gs2} - V_{t0} - \gamma\sqrt{2\Phi_F + V_{sb2}} + \gamma\sqrt{2\Phi_F})^2} \Big) \\ = \beta \cdot \Big( & \overline{(v_{gs1} - \gamma\sqrt{2\Phi_F + V_{sb1}})^2} - \overline{(v_{gs2} - \gamma\sqrt{2\Phi_F + V_{sb2}})^2} \Big) \end{aligned} \tag{5.32}$$

where $\gamma$ denotes the body effect coefficient [4], and $2\Phi_F$ is the surface potential [5]. As long as $v_{sbi}$ is much smaller than $(2\Phi_F + V_{SBi})$, a first Taylor expan-

[4] $\gamma = \sqrt{2 \cdot q \cdot \epsilon_{si} \cdot N_{sub}}/C_{ox}$ [Raz01].
[5] $\Phi_F = (kT/q) \cdot \ln(N_{sub}/n_i)$ [Raz01].

sion can be used to expand previous expression:

$$I_{OS} = \beta \cdot \left( \overline{\left( v_{gs1} - \gamma \cdot \sqrt{2\Phi_F + V_{SB1}} - \frac{\gamma}{2 \cdot \sqrt{2\Phi_F + V_{SB1}}} v_{sb1} \right)^2} - \overline{\left( v_{gs2} - \gamma \cdot \sqrt{2\Phi_F + V_{SB2}} - \frac{\gamma}{2 \cdot \sqrt{2\Phi_F + V_{SB2}}} v_{sb2} \right)^2} \right) \quad (5.33)$$

Which yields:

$$I_{OS} = \beta \cdot \left( \overline{\left( v_{gs1} - \frac{\gamma}{2 \cdot \sqrt{2\Phi_F + V_{SB1}}} v_{sb1} \right)^2} - \overline{\left( v_{gs2} - \frac{\gamma}{2 \cdot \sqrt{2\Phi_F + V_{SB2}}} v_{sb2} \right)^2} \right) \quad (5.34)$$

This expression can be rewritten using the transconductance ($g_{m1}$) and the bulk transconductance ($g_{mb1}$) [Raz01]:

$$\begin{aligned} I_{OS} &= \beta \cdot \left( \overline{\left( v_{gs1} - \frac{g_{mb1}}{g_{m1}} v_{sb1} \right)^2} - \overline{\left( v_{gs2} - \frac{g_{mb2}}{g_{m2}} v_{sb2} \right)^2} \right) \\ &= \beta \cdot \left( \overline{v_{x1}^2} - \overline{v_{x2}^2} \right) \end{aligned} \quad (5.35)$$

where $v_{xi} = v_{gsi} - \frac{g_{mbi}}{g_{mi}} v_{sbi}$. Similarly to (5.6), $V_{xi}(s)$ is expressed in terms of the common-mode and the differential-mode input voltage:

$$\begin{cases} V_{x1}(s) = H_{cm}(s) \cdot V_{cm}(s) + \frac{V_{dm}(s)}{2} \\ V_{x2}(s) = H_{cm}(s) \cdot V_{cm}(s) - \frac{V_{dm}(s)}{2} \end{cases} \quad (5.36)$$

$H_{cm}(j\omega)$ represents the common-mode transfer function. Provided that $g_{m1} = g_{m2}$, $g_{mb1} = g_{mb2}$, $g_{m3} = g_{m4}$, $C_{gs1} = C_{gs2}$, $C_{gs3} = C_{gs4}$, the common-mode transfer function $H_{cm}(s)$ is approximated by the following relation:

$$H_{cm}(s) \approx \frac{K \cdot s\left(s + \frac{2 \cdot g_{m3} \cdot C_{T1} \cdot (g_{m1} + g_{mb1})}{K}\right)}{A \cdot \left(s + \frac{2 \cdot g_{m3}}{C_{T2} + 2 \cdot C_{gs3}}\right) \cdot \left(s + \frac{2 \cdot (g_{m1} + g_{mb1})}{C_{T1} + 2 \cdot C_{gs1}}\right)} \quad (5.37)$$

where $A$ and $K$ are respectively equal to:

$$\begin{cases} A = g_{m1} \cdot (C_{T1} + 2 \cdot C_{gs1}) \cdot (C_{T2} + 2 \cdot C_{gs3}) \\ K = g_{m1} \cdot C_{T1} \cdot C_{T2} - 2 \cdot g_{mb1} \cdot C_{gs1} \cdot C_{T2} + 2 \cdot (g_{m1} + g_{mb1}) \cdot C_{T1} \cdot C_{gs3} \\ \qquad + 2 \cdot g_{m1} \cdot C_{bs1} \cdot (C_{T1} + C_{T2}) \end{cases}$$

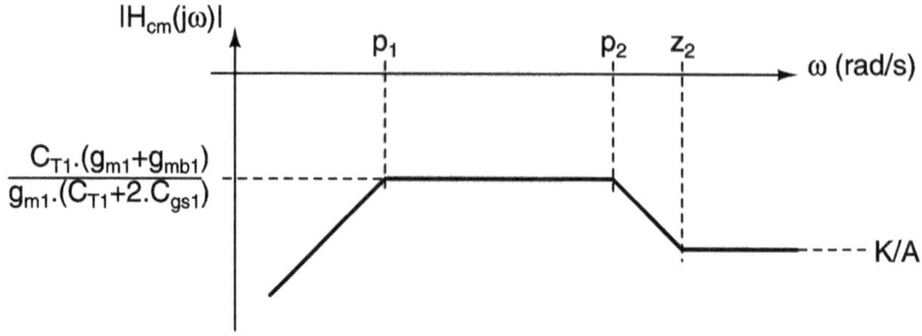


*Figure 5.13.* Simplified Bode plot of $|H_{cm}(j\omega)|$ of the source-buffered differential pair.

In order to minimize the input offset voltage, $|H_{cm}(s)|$ must be reduced as much as possible, as was illustrated anteriorly. This can be realized by increasing $C_{gs1}$ and by setting $K$ to zero, hereby warping the second zero of $H_{cm}(s)$ to very high frequencies. For $K$ to be equal to zero, $C_{gs1}$ must be equal to:

$$C_{gs1} = \frac{g_{m1} \cdot C_{T1} \cdot C_{T2} + 2 \cdot g_{m1} \cdot C_{bs1} \cdot (C_{T1} + C_{T2}) + 2 \cdot C_{T1} \cdot C_{gs3} \cdot (g_{m1} + g_{mb1})}{2 \cdot C_{T2} \cdot g_{mb1}} \tag{5.38}$$

Two small on-chip gate-source capacitances ($C_{in}$) are added between the gates and the sources of $M_1$ and $M_2$ in order to comply with relation (5.38). A large EMI induced offset cancellation is achieved this way, as was illustrated in [Red06c].

A larger total gate-source capacitance $C_{gs1}$ decreases the maximal level of $|H_{cm}(s)|$, as shown in the simplified Bode plot on Fig. 5.13. Observe that the maximum level of this transfer function is much smaller than the one obtained for the classic differential pair (Fig. 5.5), since $C_{T1}$ is much smaller (it does not contain the parasitic bulk-well capacitance any longer, and solely consists of the parasitic drain-bulk capacitance of $M_{b1}$) while $C_{gs1}$ is in turn much larger (it has been increased with $C_{in}$) in order to comply with (5.38). This constitutes a very important benefit when dealing with high input signals, since this means that the slew rate asymmetry generated by the input differential pair is significantly reduced (refer to Sect. 2.1), while the linearity when dealing with large input signals is greatly improved as well (refer to Sect. 2.2). An extra advantage of this circuit is the fact that the input referred noise of this circuit equals the input noise of a classic differential pair, since $M_3$ and $M_4$ do not perturb the signal path of $M_1$-$M_2$. Therefore:

$$\overline{V_{n,in}{}^2} = 8kT \cdot \left(\frac{2}{3 \cdot g_{m1}}\right) \tag{5.39}$$

This minimal input noise represents a huge improvement compared to previous circuits [6]. However, this circuit solution is not without flaws: the main disadvantage of the source-buffered differential pair resides in the sensitivity to systematic offset of capacitors $C_{in}$. Indeed, when integrated, $C_{in}$ exhibits a non-negligible systematic error, resulting in a non zero $K$ term. This is illustrated in the following example.

*Example*: Consider a source-buffered differential pair with $g_{m1} = g_{m2} = g_{m3} = g_{m4} = 710$ µS, $C_{gs1} = C_{gs2} = C_{gs3} = C_{gs4} = 117$ fF, $C_{T1} = 68.5$ fF, $C_{T2} = 529$ fF. Two extra capacitors $C_{in}$ equal to 359 fF have been connected between the gates and the sources of $M_1$ and $M_2$ in order to make $K$ zero (refer to (5.37)). The resulting output offset has been compared to the offset generated by a classic differential pair dimensioned as in the example in Sect. 2.3.2, and plotted in Fig. 5.14: observe that the resulting EMI induced input offset is very small. The injected EMI signal level equals $V_d = 200$ mV RMS and $V_{cm} = 100$ mV RMS as in the previous examples. However, owing to the systematic offset of $C_{in}$, $K$ is not perfectly equal to zero. This results in an increased offset owing to the systematic error of $C_{in}$. This dependence is illustrated in Fig. 5.15: as observed in this plot, the dependence to the systematic offset of $C_{in}$ is considerable.

As illustrated in the previous example, the major disadvantage of the source-buffered differential pair is its high dependency on tightly specified tolerances of on-chip capacitors. The latter display considerable variability in the order of 20%–30%, mostly due to process variations [Has01]. As was demonstrated previously, the total $C_{gs1}$ must equal (5.38) in order to make $K$ zero. This is realized by placing two extra capacitors in parallel with $C_{gs1}$ and $C_{gs2}$ ($C_{in}$ in Fig. 5.12). However, any variation in $C_{in}$ translates itself in a non zero $K$, which increases the EMI induced offset voltage. This issue is overcome in the present circuit, depicted in Fig. 5.16, in which a source resistance has been added. Analogously to the derivation presented in the previous sections,

---

[6] This observation is, however, not so trivial as it may look like. The auxiliary differential pair generates noise at the sources of $M_3$ and $M_4$, and this noise is injected into the bulk of $M_1$-$M_2$. The noise voltage at the sources of $M_3$ and $M_4$ is expressed as:

$$\overline{V_{n,x}{}^2} = 8kT \cdot \left(\frac{2}{3 \cdot g_{m3}}\right) \tag{5.40}$$

Referring the total noise to the input of $M_1$-$M_2$ yields:

$$\overline{V_{n,in}{}^2} = 8kT \cdot \left(\frac{2}{3 \cdot g_{m3}} \cdot \frac{g_{mb1}^2}{g_{m1}^2} + \frac{2}{3 \cdot g_{m1}}\right) \tag{5.41}$$

The ratio $g_{mb1}/g_{m1}$ is typically small, and equal to $C_D/C_{ox}$, where $C_D$ and $C_{ox}$ are respectively the channel-bulk and the oxide-channel capacitance. Usually, $C_D$ is around 1/3 of $C_{ox}$, as given in [San06]. Since $g_{m3}$ is chosen equal to $g_{m1}$, the term $\frac{2}{3 \cdot g_{m3}} \cdot \frac{g_{mb1}^2}{g_{m1}^2}$ is negligible compared to $\frac{2}{3 \cdot g_{m1}}$.

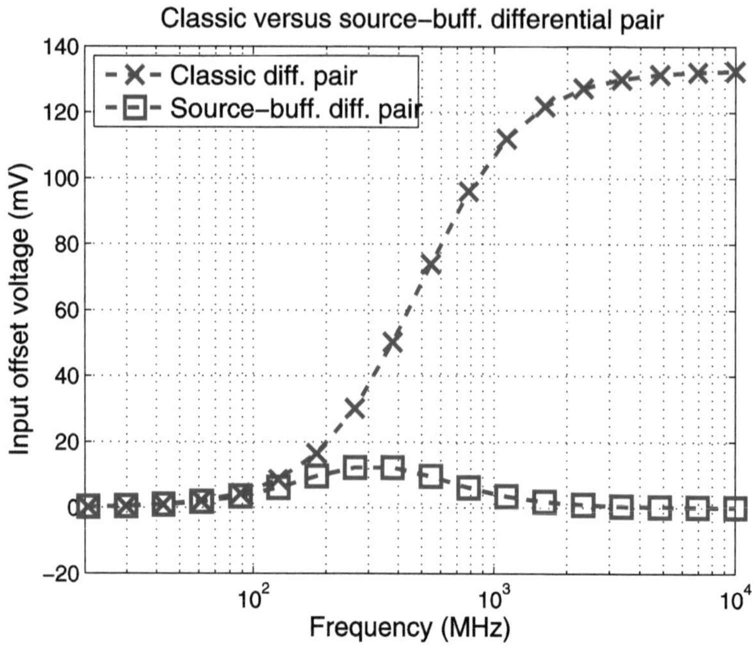


*Figure 5.14.* Input referred offset voltages of the source-buffered differential pair and of the classic differential pair plotted against frequency.

$V_{xi}(j\omega)$ is expressed in terms of the common-mode and the differential-mode voltage:

$$\begin{cases} V_{x1}(j\omega) = H_{cm}(j\omega) \cdot V_{cm}(j\omega) + H_{dm}(j\omega) \cdot \frac{V_{dm}(j\omega)}{2} \\ V_{x2}(j\omega) = H_{cm}(j\omega) \cdot V_{cm}(j\omega) - H_{dm}(j\omega) \cdot \frac{V_{dm}(j\omega)}{2} \end{cases} \tag{5.42}$$

$H_{cm}(j\omega)$ and $H_{dm}(j\omega)$ represent the common-mode and the differential-mode transfer function. $H_{cm}(j\omega)$ is identical to (5.37), since the common-mode transfer function is unchanged. Provided that all transistors are identical as well as identically biased ($g_{m1} = g_{m2}$, $g_{mb1} = g_{mb2}$, $g_{m3} = g_{m4}$, $C_{gs1} = C_{gs2}$, $C_{gs3} = C_{gs4}$), $H_{dm}(j\omega)$ is expressed as follows:

$$H_{dm}(j\omega) = \frac{\left(C_{T1} \cdot R_o + 2 \cdot R_s \cdot C_{bs1} - 2 \cdot R_s \cdot (C_{gs1} + C_{in}) \cdot \frac{g_{mb1}}{g_{m1}}\right) \cdot s + 4}{(C_{T1} \cdot R_s + 2 \cdot R_s \cdot C_{bs1} + 2 \cdot R_s \cdot (C_{gs1} + C_{in})) \cdot s + 2 \cdot R_s \cdot (g_{m1} + g_{mb1}) + 4} \tag{5.43}$$

The input offset is now equal to:

$$V_{OS} = \frac{1}{V_{GS1} - V_t} \int_{-\infty}^{\infty} \Big| H_{cm}(j\omega) \cdot V_{cm}(j\omega) \cdot H_{dm}(j\omega) \cdot V_{dm}(j\omega) \Big| \cdot \cos\phi \cdot d\omega \tag{5.44}$$

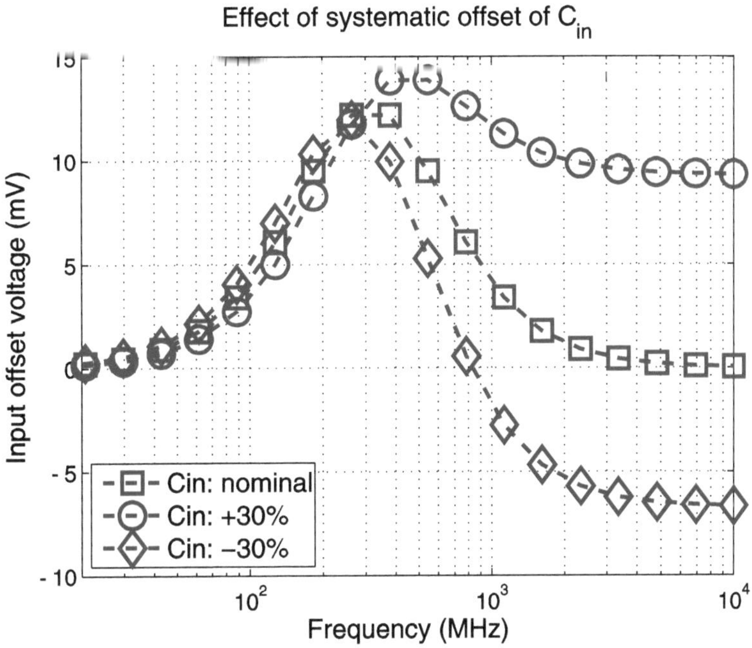


*Figure 5.15.* Input referred offset voltage of the source-buffered differential pair plotted against frequency, for different systematic offsets of $C_{in}$.

where:

$$\phi = \operatorname{atan}\left(\frac{\operatorname{Im}\{H_{cm}(j\omega)\cdot V_{cm}(j\omega)\cdot H_{dm}(j\omega)\cdot V_{dm}(j\omega)\}}{\operatorname{Re}\{H_{cm}(j\omega)\cdot V_{cm}(j\omega)\cdot H_{dm}(j\omega)\cdot V_{dm}(j\omega)\}}\right)$$

In order to minimize the input offset voltage, $|H_{cm}(j\omega)|$ must be reduced as much as possible, as was described in the previous paragraph. However, as stated previously, integrated capacitors are subject to systematic offset: such variations significantly alter the total value of $C_{gs1}$, which increases the input offset voltage because $K$ is no longer equal to zero. In order to counter this effect, resistor $R_s$ is connected between the sources of $M_1$ and $M_2$, which linearizes the input transconductance $g_{m1}$: this alleviates the EMI induced offset caused by the nonlinearity of the differential pair. More importantly, the differential-mode transfer function $H_{dm}(j\omega)$ becomes very small at high frequencies, because the sources of $M_1$ and $M_2$ are no longer constituting a differential ground. This yields:

$$H_{dm}(\infty) = \frac{C_{T1} + 2\cdot C_{bs1} - 2\cdot (C_{gs1} + C_{in})\frac{g_{mb1}}{g_{m1}}}{(2\cdot C_{gs1} + C_{in}) + 2\cdot C_{bs1} + C_{T1}} \tag{5.45}$$

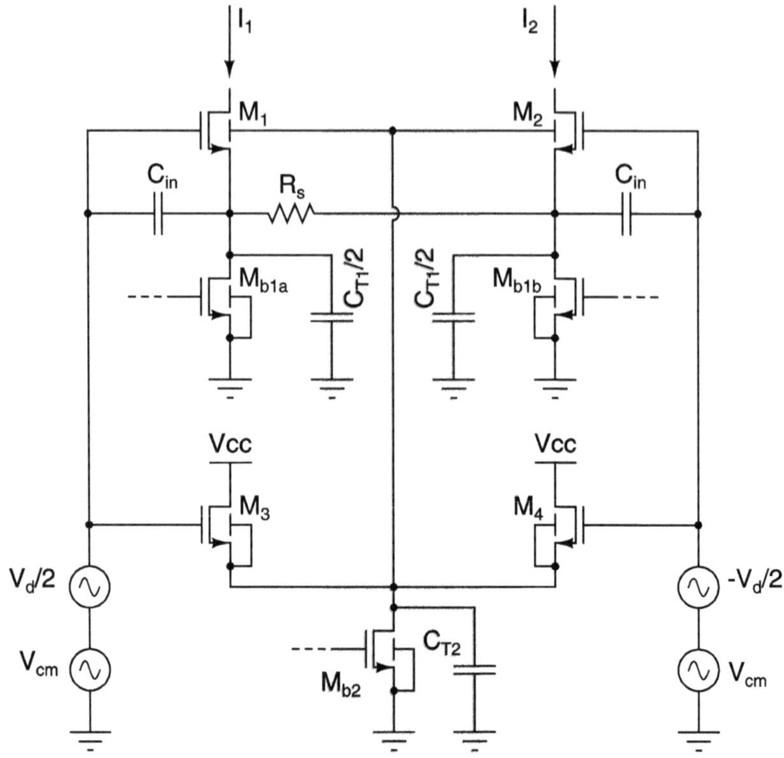


*Figure 5.16.* Source-buffered differential pair with source resistance $R_s$, disturbed by a common-mode and a differential-mode EMI signal of the same frequency.

The resulting input offset voltage at high EMI frequencies is expressed as:

$$V_{OS}(\infty) = \frac{1}{V_{GS1} - V_t} \int_{-\infty}^{\infty} \left| \frac{K}{A} \right| \cdot \left| \frac{C_{T1} + 2 \cdot C_{bs1} - 2 \cdot (C_{gs1} + C_{in}) \frac{g_{mb1}}{g_{m1}}}{2 \cdot (C_{gs1} + C_{in}) + 2 \cdot C_{bs1} + C_{T1}} \cdot V_{cm}(j\omega) \cdot V_{dm}(j\omega) \right| \cdot \cos\phi \cdot d\omega \quad (5.46)$$

Provided that $C_{in}$ is much larger than $C_{bs1}$ and $C_{T1}$, previous expression simplifies to:

$$V_{OS}(\infty) = \frac{1}{V_{GS1} - V_t} \int_{-\infty}^{\infty} \left| \frac{K}{A} \right| \cdot \left| \frac{g_{mb1}}{g_{m1}} \cdot V_{cm}(j\omega) \cdot V_{dm}(j\omega) \right| \cdot \cos\phi \cdot d\omega \quad (5.47)$$

Observe that since $H_{dm}(\infty)$ is equal to $g_{mb1}/g_{m1}$, the impact of the term $K/A$ is reduced at higher frequencies by a factor roughly equal to 3 or 4. Without $R_s$, $|H_{dm}(j\omega)|$ would be equal to 1, which would in turn increase the contribution of $K$. The input referred noise is larger however, because of the presence of source resistance $R_s$: both $M_{b1a}$ and $M_{b1b}$ contribute to the differential noise, and this means that the total input noise increases:

$$\overline{V_{n,in}}^2 = 8kT \cdot \left( \frac{2}{3 \cdot g_{m1}} + \frac{g_{m3} \cdot R_s^2}{6} + \frac{R_s}{2} \right) \tag{5.48}$$

This is however not as much as it may seem, since the degeneration resistor is usually not larger than $1/g_{m1}$. Assuming the worst case situation when $1/g_{m1} = R_s$, expression (5.48) simplifies to:

$$\begin{aligned} \overline{V_{n,in}}^2 &= 8kT \cdot \left( \frac{2}{3 \cdot g_{m1}} + \frac{1}{6 \cdot g_{m1}} + \frac{1}{2 \cdot g_{m1}} \right) \\ &= 8kT \cdot \left( \frac{4}{3 \cdot g_{m1}} \right) \end{aligned} \tag{5.49}$$

Which is simply twice as much as the equivalent input noise of a classic differential pair expressed in (5.15).

*Example*: Consider a source-buffered differential pair with $g_{m1} = g_{m2} = g_{m3} = g_{m4} = 710\ \mu\text{S}$, $C_{gs1} = C_{gs2} = C_{gs3} = C_{gs4} = 117$ fF, $C_{T1} = 68.5$ fF, $C_{T2} = 529$ fF. Two extra capacitors $C_{in}$ equal to 359 fF have been connected between the gates and the sources of $M_1$ and $M_2$ in order to make $K$ zero (refer to 5.37), as previously. The degeneration resistor $R_s$ has been chosen equal to 1200 Ω. Figure 5.17 depicts the effect of the systematic offset of $C_{in}$. Compared to the plot depicting the offset voltage obtained without $R_s$ (Fig. 5.15), the improvement is significant.

### 2.3.7 Comparison

Based on previous calculations, the source-buffered differential pair using a small resistive source degeneration clearly offers the best alternative to reduce the nonlinearity of the input stage, in terms of smallest EMI induced offset voltage and highest insensitivity to process variations (like mismatch and systematic offset). The total input referred noise is also quite attractive compared to other solutions, as long as the source degeneration resistor is not too large: Table 5.1 presents an overview, listing the input referred noise as well as the required bias current of each discussed differential pair structure.

## 2.4 EMI induced offset measurement setups

As was derived previously for weak nonlinear behavior, the EMI induced offset in a differential pair is proportional to the scalar product of the differential and common-mode component of the EMI signal that is injected in the input terminals of the opamp. Since the EMI induced worst-case offset voltage is maximal when the differential and common-mode EMI components at the differential

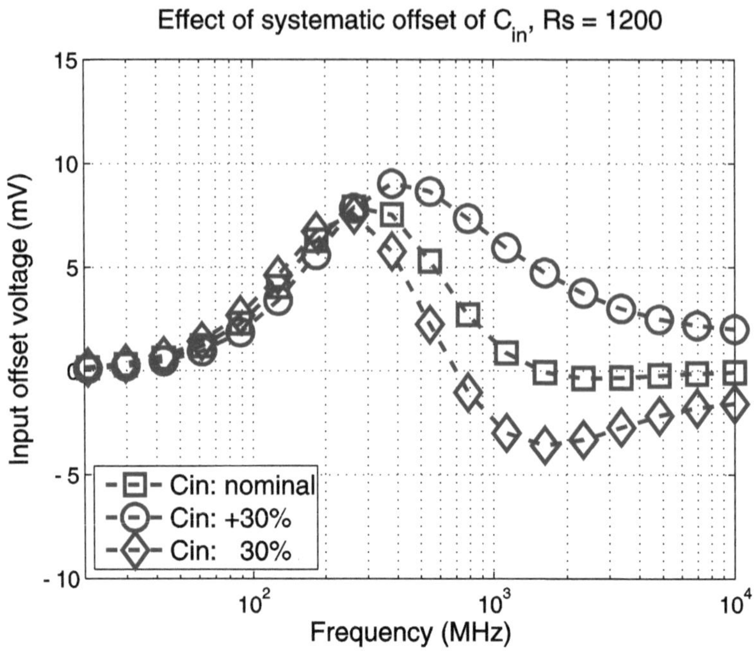


*Figure 5.17.* Input referred offset voltage of the source-buffered differential pair plotted against frequency, for different systematic offsets of $C_{in}$.

pair inputs are non zero as well as in phase with each other (refer to (5.8)), an ideal measurement setup should reproduce this situation in order to ascertain the resulting worst-case offset voltage. This section studies and compares possible opamp measurement topologies for determining the EMI induced input offset voltage when EMI is injected into the input terminals of the opamp in question. The ideal measurement setup should allow a correct EMI induced offset voltage measurement, and this preferably for a broad range of EMI frequencies. Furthermore, it should be easy to use and yield correct measurement results up to very high EMI frequencies (up to 1 GHz, according to the current EMI measurement standards, like [DPI]).

- **Voltage follower:** In most measurement setups, the EMI induced input offset is measured by connecting the opamp as a voltage follower and measuring the DC level at the output (Fig. 5.18a) [Ric01, Fio01, Fio02a].

  Consider that the opamp is modeled as a perfect one pole system. Its open loop gain function can then be expressed as follows:

$$A(s) = \frac{A_{DC}}{1 + \frac{s}{p_1}} \tag{5.50}$$

*Table 5.1.* Comparison of differential pairs in terms of current consumption and noise.

| | *Schematic* | *Equivalent input noise* $\overline{V_{n,in}}^2$ $(V^2/Hz)$ | *Current consumption* |
|---|---|---|---|
| Classic differential pair | Fig. 5.3 | $8kT \cdot \left(\frac{2}{3 \cdot g_{m1}}\right)$ | $I_{tot} = I_b$ |
| Classic differential pair with resistive source degeneration | Fig. 5.6 | $8kT \cdot \left(\frac{2}{3 \cdot g_{m1}} + R_s\right)$ | $I_b < I_{tot}$ |
| Cross-coupled differential pair with high-pass $R$-$C$ | Fig. 5.8 | $8kT \cdot \left(\frac{4}{3 \cdot g_{m1}} + R_1\right)$ | $I_{tot} = 2 \cdot I_b$ |
| Classic differential pair with low-pass $R$-$C$ | Fig. 5.10 | $8kT \cdot \left(\frac{2}{3 \cdot g_{m1}} + R_1\right)$ | $I_{tot} = I_b$ |
| Improved cross-coupled differential pair | Fig. 5.11 | $8kT \cdot \left(\frac{2}{3 \cdot g_{m1}} + \frac{2}{3 \cdot g_{m3}}\right)$ | $I_b < I_{tot} \leq 2 \cdot I_b$ |
| Source-buffered differential pair | Fig. 5.12 | $8kT \cdot \left(\frac{2}{3 \cdot g_{m1}}\right)$ | $I_b < I_{tot}$ |
| Source-buffered differential pair with degenerated source | Fig. 5.16 | $8kT \cdot \left(\frac{2}{3 \cdot g_{m1}} + \frac{2 \cdot g_{m3} \cdot R_s^2}{3} + \frac{R_s}{2}\right)$ | $I_b < I_{tot}$ |

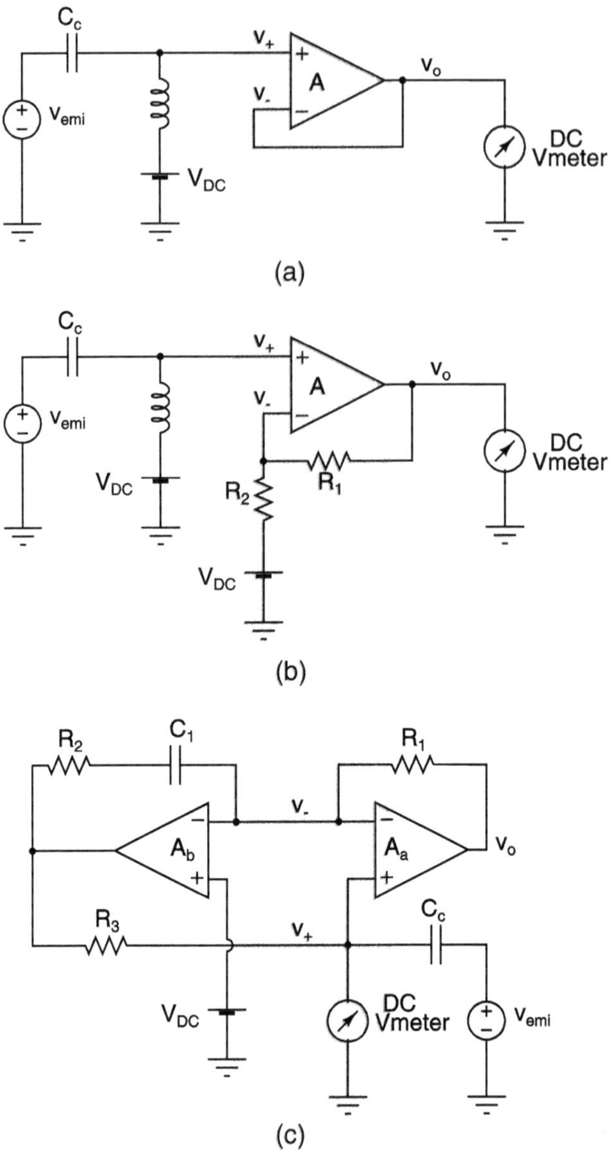


*Figure 5.18.* EMI induced DC shift opamp measurement setups (a) Voltage follower. (b) Non-inverting amplifer. (c) Double opamp measurement structure.

where $A_{DC}$ represents the DC voltage gain and $p_1$ the dominant opamp pole. Provided that the DC gain is large, the output voltage is derived as follows:

$$v_o(s) = \frac{A_{DC} \cdot v_{emi}(s)}{\frac{s}{p_1} + A_{DC}} \tag{5.51}$$

The signal appearing at the inverting opamp input is equal to the output voltage owing to the negative feedback loop ($V_-(t) = V_o(t)$). The input referred offset voltage is consequently determined by measuring the DC level at the opamp output. Two separate cases can be distinguished:

- For EMI frequencies lying significantly higher than the gain bandwidth product (GBW $= p_1 \cdot A_{DC}$) of the operational amplifier, the AC signal at the output of the opamp is close to zero and consequently the common and differential-mode signals at the opamp input terminals are expressed as:

$$\begin{aligned} V_{dm}(t) &= v_{emi}(t) \\ V_{cm}(t) &= V_{DC} + \frac{v_{emi}(t)}{2} \end{aligned} \tag{5.52}$$

- For EMI frequencies close to and below the GBW, the differential-mode component is heavily attenuated owing to the negative feedback loop:

$$\begin{aligned} V_{dm}(t) &\approx 0 \\ V_{cm}(t) &= V_{DC} + v_{emi}(t) \end{aligned} \tag{5.53}$$

As can be observed in (5.52) and (5.53), for EMI frequencies close to and below the GBW frequency, the voltage follower measurement setup yields a very small offset voltage which does not correspond to the worst-case offset voltage any longer.

- **Non-inverting amplifier structure:** The differential-mode component can be increased by decreasing the AC feedback; the non-inverting amplifier structure depicted in Fig. 5.18b allows measurement of the worst case EMI induced voltage offset for lower EMI frequencies. Provided that $A_{DC}$ is large, the output voltage can be approximated by:

$$v_o(s) = \frac{A_{DC} \cdot v_{emi}(s)}{\frac{s}{p_1} + (\frac{R_2}{R_1+R_2}) \cdot A_{DC}} \tag{5.54}$$

The signal appearing on the inverting input is now expressed as:

$$v_-(t) = \frac{R_2}{R_1 + R_2} \cdot v_o(t) \tag{5.55}$$

In principle, as long as the lowest EMI frequencies are higher than the $GBW$ times $\frac{R_2}{R_1+R_2}$, the common and differential-mode components are equal to (5.52), which means that lower EMI frequencies can be measured depending on the resistive divider formed by $R_1$ and $R_2$. Unfortunately, both resistors equally interfere with the DC loop of the circuit and therefore cause the opamp output to saturate for disproportionate values. This limits the value of the lowest measurable EMI frequencies. Replacing $R_2$ with a capacitor reduces the AC feedback while preserving the DC feedback, but creates an extra pole which may impair the stability of the feedback loop.

- **Double opamp measuring structure:** The proposed double opamp structure circumvents the aforementioned imperfections by separating the DC feedback from the AC feedback loop and by using a Miller integrator to filter the AC loop [Red07d]. Refer to Fig. 5.18c. This structure requires two opamps, $A_a$ and $A_b$: $A_a$ is the opamp whose offset is measured whereas $A_b$ forms the Miller integrator and completes the DC feedback loop. Since two opamps are now present in the feedback loop, the stability of the full circuit must be ascertained, and this for all possible conditions (whether or not EMI is injected in the circuit and whether or not the external coupling capacitor ($C_c$) is connected to it). These two cases are considered separately:

  - $C_c$ is connected to the circuit: stability is ensured since $R_3$ and $C_c$ are very large ($R_3$ limits the EMI coupling into the output of $A_b$ while $C_c$ is typically equal to 6.8 nF as specified in [DPI]). Consequently, $R_3$ and $C_c$ form a dominant pole which lies at very low frequencies.
  - $C_c$ is not connected to the circuit: since there is no large dominant pole at the non-inverting input of $A_a$, the stability may be compromised. This is now checked in detail. In the event that $A_a$ equals $A_b$ (both opamps are equal to each other) and that $R_2$ is small compared to $R_1$, the open loop transfer function from the non-inverting input of $A_a$ to the output of (Fig. 5.19) contains 3 poles, and one zero:

$$A_{OL}(s) = \frac{v_{out}(s)}{v_{emi}(s)} = -\frac{A_{DC} \cdot (R_2 \cdot C_1 \cdot s + 1)}{(\frac{s}{p_1} + 1) \cdot (\frac{s}{A_{DC} \cdot p_1} + 1) \cdot (R_1 \cdot C_1 \cdot s + 1)} \tag{5.56}$$

    The circuit is typically dimensioned so that $1/(R_1 \cdot C_1)$ forms the dominant pole. $R_2$ forms a zero with $C_1$: this zero is chosen to lie at the same frequency as $p_1$, and as such cancels out this pole. The remaining two pole system is stable, as long as $1/(R_1 \cdot C_1)$ lies at a lower frequency than the dominant pole $p_1$ [7].

Now that the stability is ensured, the output voltage of $A_a$ ($v_o(s)$) is easily calculated using the same small signal equivalent and observing the same

[7] In case $\frac{1}{R_1 \cdot C_1} = p_1$, the phase margin is equal to 45°.

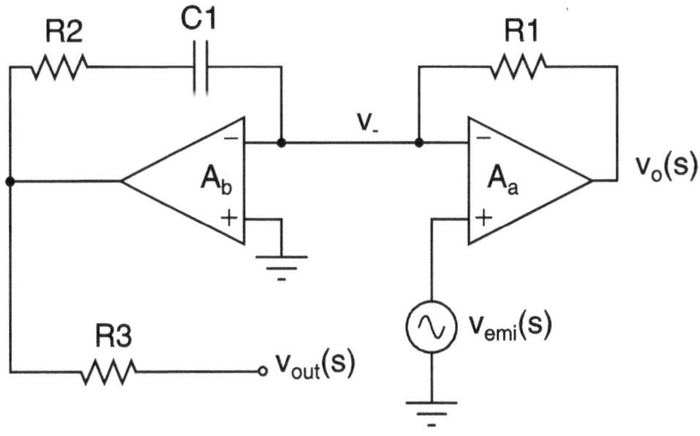


*Figure 5.19.* Double opamp measurement structure: AC open loop schematic.

conditions as previously:

$$\frac{v_o(s)}{v_{emi}(s)} = \frac{\frac{C_1 \cdot R_1 \cdot s^2}{p_1} + C_1 \cdot R_1 \cdot A_{DC} \cdot s + 1}{(\frac{s}{p_1} + 1) \cdot (C_1 \cdot R_1 \cdot s + 1) \cdot (\frac{s}{A_{DC} \cdot p_1} + 1)} \tag{5.57}$$

Under normal circumstances, $A_{DC}$ is large, and previous transfer function can be simplified to:

$$\frac{v_o(s)}{v_{emi}(s)} \approx \frac{C_1 \cdot R_1 \cdot A_{DC} \cdot s}{(\frac{s}{p_1} + 1) \cdot (C_1 \cdot R_1 \cdot s + 1)} \tag{5.58}$$

The signal voltage at the non-inverting input of $A_a$ is equally easily calculated:

$$v_{-}(s) \approx \frac{p_1 \cdot A_{DC} \cdot (R_2 \cdot C_1 \cdot s + 1) \cdot v_{emi}(s)}{(R_1 \cdot C_1 \cdot s + 1) \cdot (s + p_1 \cdot A_{DC})} \tag{5.59}$$

As can be observed, the signal at the inverting input of $A_a$ becomes negligible for EMI frequencies which are superior to $1/(R_1 \cdot C_1)$, causing the common-mode and the differential-mode component at the input terminals to equal (5.52). Since $1/(R_1 \cdot C_1)$ can be made quite small, this EMI induced offset measurement structure yields a considerable improvement compared to the previous structures. Furthermore, the lowest measurable EMI frequency is now fully independent of the opamp GBW and of the opamp characteristics. The input referred offset is easily found by measuring the DC voltage shift at the non-inverting input of $A_a$. A small extra filtering capacitor between $v_{-}$ and ground can be foreseen internally.

The performances of the three discussed measurement topologies are compared with their respective Bode plots in Fig. 5.20. Observe that $|A_{OL}(\omega)|$ represents the open loop gain transfer function, and $\delta$ the logarithmic difference

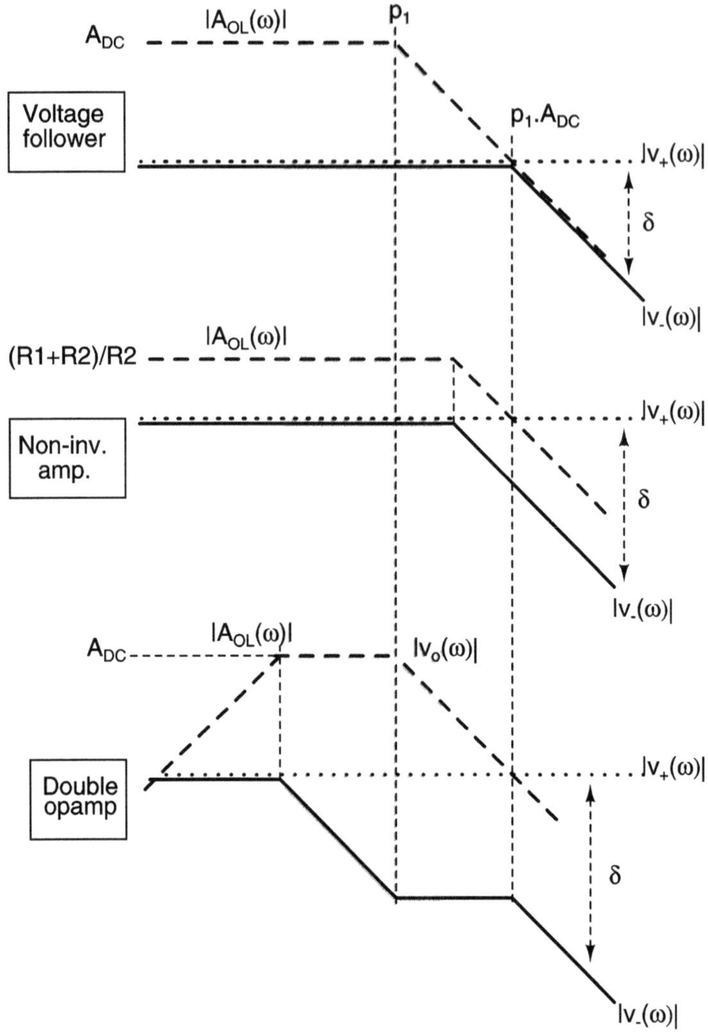


*Figure 5.20.* Bode plots of the voltage follower, non-inverting amplifier and double opamp measurement topologies.

between the EMI source $v_{emi}(j\omega)$ injected into $v_+(j\omega)$ and $v_-(j\omega)$. Clearly, for a large $\delta$, the EMI signal components at the opamp inputs are expressed as in (5.52). The complementary statement is true for a small $\delta$: in that case, $v_+(j\omega)$ and $v_-(j\omega)$ are almost equal to each other, meaning that the differential EMI component is close to zero, as is expressed mathematically in (5.53).

*Example*: The performances of the three discussed measurement topologies have been compared with simulations using a test case. The opamp has been

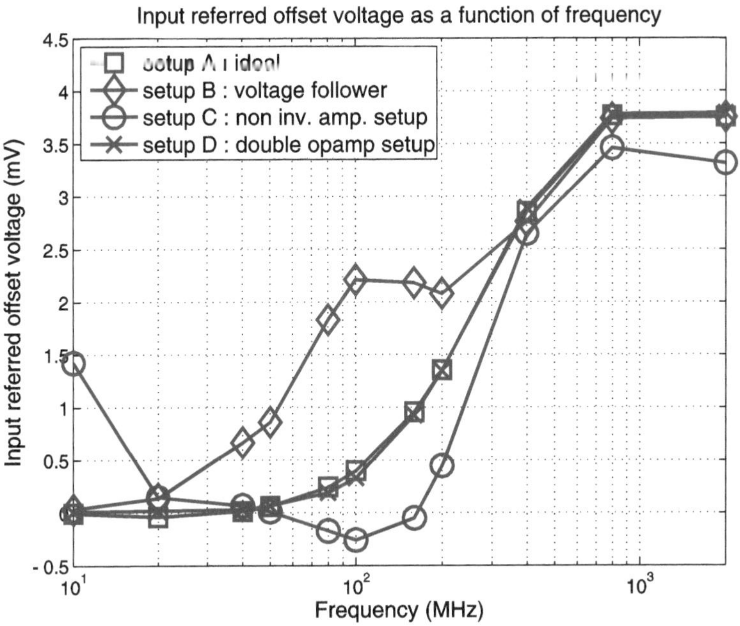


*Figure 5.21.* Comparing the different measurement topologies: input referred offset voltage against frequency.

designed as a folded cascode topology with a GBW of 77 MHz and a DC gain of 45 dB in a 0.35 μm CMOS technology. The EMI induced offset of this opamp resulting from the three described measurement setups is plotted in Fig. 5.21, along with the expected theoretical offset value. The amplitude of the EMI disturbance is equal to 50 mV. As can be seen in Fig. 5.21, the voltage follower configuration as well as the non-inverting amplifier setup yield correct results starting from approximately 300 MHz. Clearly, the double opamp measurement configuration yields the best results: the EMI induced offset closely follows the theoretical curve, and this for all simulated EMI frequencies.

## 2.5 Measurements

A test-IC containing four different circuits has been processed in the AMIS 0.35 μm CMOS technology [8]. A detailed floorplan as well as a microphotograph of the circuit are shown respectively in Figs. 5.22 and 5.23. The classic differential pair and the source-buffered differential pair with resistive degeneration have been implemented in the designs of two otherwise identical folded cascode opamps [Raz01], labeled respectively $A_1$ and $A_2$: the design specifi-

[8] Processed in AMIS I3T80, using the low-voltage components [Moe02].

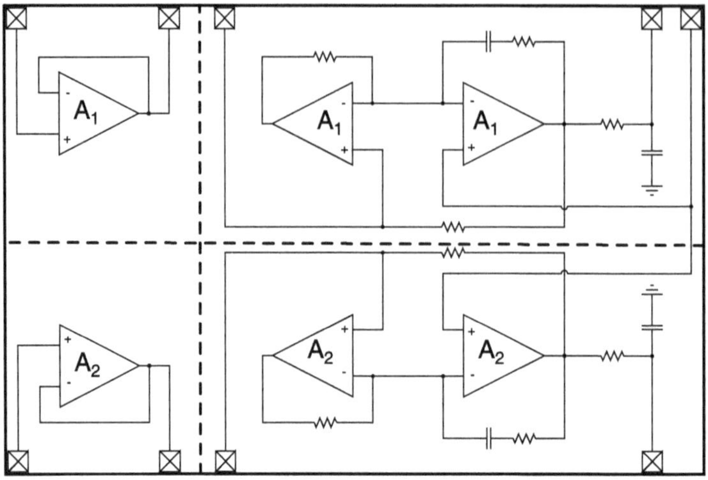


*Figure 5.22.* Floorplan of the test-IC. $A_1$ and $A_2$ are folded cascode opamps, similar to each other except for the input differential pair: $A_1$ contains the classic differential pair, $A_2$ the source-buffered one (with $R_s$). Both opamps are measured with the voltage follower configuration as well as with the double opamp measurement setup.

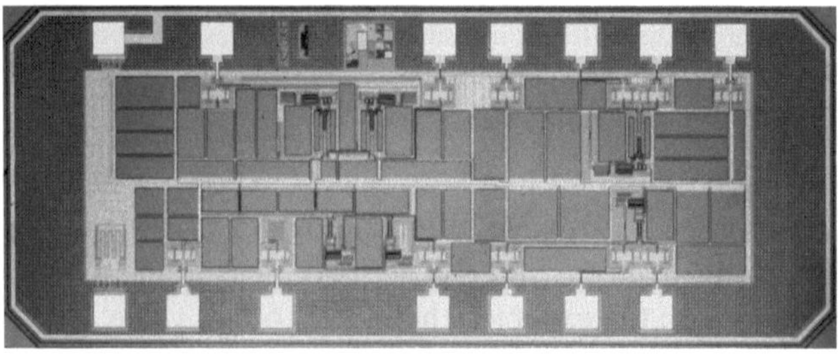

*Figure 5.23.* Microphotograph of the test-IC. The size of the circuit is approximately 1.9 mm × 0.9 mm.

cations are enumerated in Table 5.2. The purpose of both opamps is to form a framework in which the EMI behavior of the classic differential pair versus the source-buffered differential pair is measured and compared.

- **Low EMI frequencies (below 200 MHz):** For small frequencies, the voltage follower measurement configuration can not be used to measure the correct EMI induced offset voltage as has been derived in Sect. 2.4. Figure 5.24 shows the EMI induced offset as a function of the EMI frequency when an input EMI signal of 50 mV RMS is applied at the inputs of both folded cascode opamps (containing respectively a classic and a source-buffered differential input pair) using the double opamp measurement setup: observe that the measured data fits the theoretical curve.

*Table 5.3.* Design specifications.

| | $A_1$ *(classic diff. pair)* | $A_2$ *(source-buff. diff. pair with $R_s$)* |
|---|---|---|
| $A_{DC}$ | 54 dB | 51 dB |
| GBW | 36 MHz | 24 MHz |
| $g_{m1}$ | 710 μS | 710 μS |
| $R_s$ | – | 1.2 kΩ |
| Power (including output buffers) | 3.2 mW | 3.5 mW |

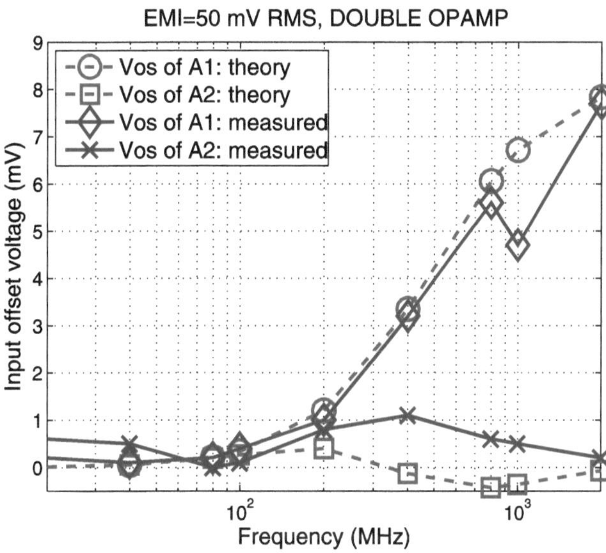


*Figure 5.24.* Measurements using the double opamp configuration (injected EMI level = 50 mV RMS).

Figure 5.25 shows the EMI induced offset for an input EMI signal of 750 mV RMS using the double opamp measurement setup. This figure illustrates that the output of $A_1$ (containing the classic differential pair) is heavily saturated, and confirms that the EMI induced offset progressively drops to zero for low EMI frequencies as predicted by the theoretical first order model.

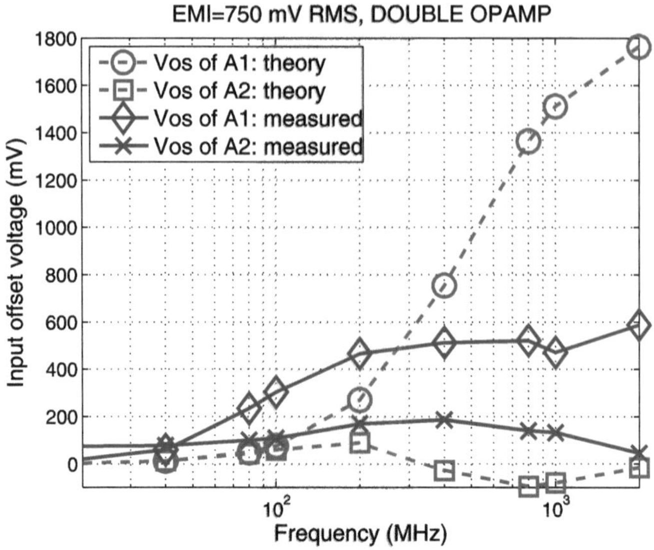


*Figure 5.25.* Measurements using the double opamp configuration (injected EMI level = 750 mV RMS).

- **High EMI frequencies (above 200 MHz):** for higher EMI frequencies, the voltage follower measurement configuration is used. Figure 5.26 shows the EMI induced offset for a small EMI signal of 50 mV RMS. Note the close correspondence between the theoretical curve and the measured curve starting from 200 MHz up to 800 MHz. For EMI frequencies lying below 200 MHz, the voltage follower configuration yields inaccurate results as predicted in Sect. 2.4. For EMI frequencies exceeding 800 MHz, the parasitic effects of the PCB tracks, the bond wires and the ESD protections cause the measured curve to diverge from the theoretical first order model. Note that the offset of $A_2$ is higher than what the theoretical model predicts: this stems from the approximations used in the first order model, as well as from inaccuracies in the modeling of the well capacitancess. Figure 5.27 shows the EMI induced offset for a large EMI input signal of 750 mV RMS. The maximal EMI induced input offset voltage corresponds to respectively 116 mV for $A_2$ (containing the proposed source-buffered topology) compared to 610 mV for $A_1$ (with the classical differential pair). Observe that the measured EMI induced offset of $A_1$ diverges from the theoretical curve because the output of the opamp is strongly saturated, meaning that the observed offset reduction achieved with the source-buffered differential pair is even higher than what the figures derived from Fig. 5.27 indicate.

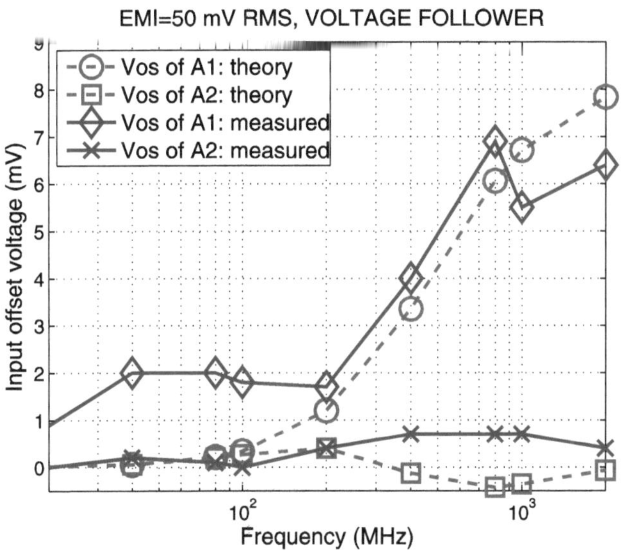


*Figure 5.26.* Measurements using the voltage follower configuration (injected EMI level = 50 mV RMS).

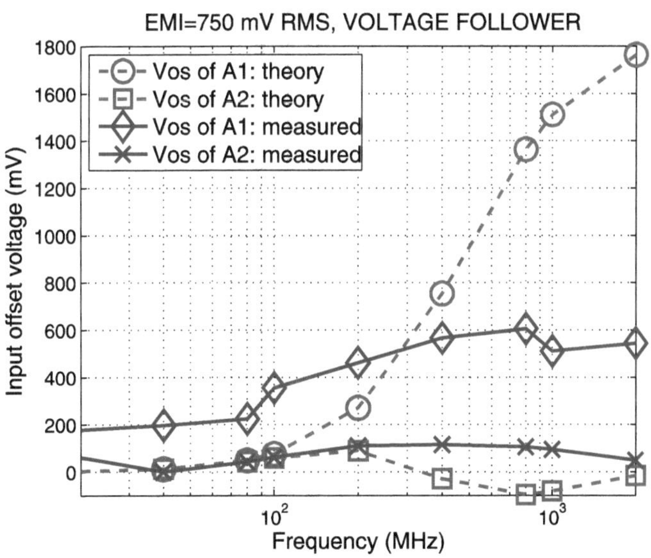


*Figure 5.27.* Measurements using the voltage follower configuration (injected EMI level = 750 mV RMS).

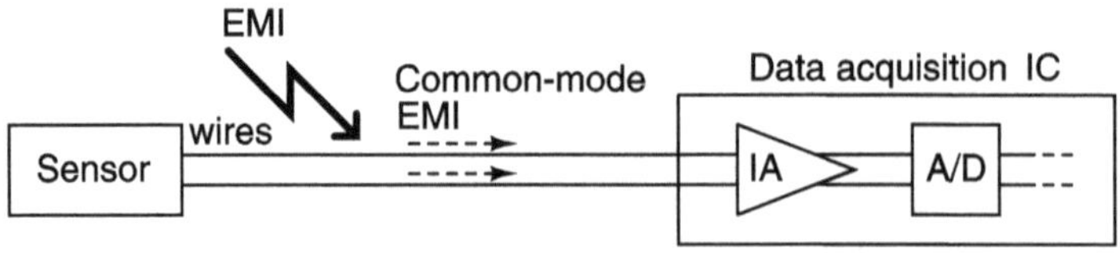


*Figure 5.28.* Typical drive-by-wire network. The sensor is wired to the data acquisition IC comprised of an instrumentation amplifier (IA) followed by an A to D converter (A/D). The noisy electromagnetic environment generates common-mode EMI in the wire pair.

## 3 Case study 2: EMI resisting instrumentation amplifier input circuit

Automotive drive-by-wire and fly-by-wire control systems in which actuators and sensor networks are connected to data acquisition systems by means of (long) electric wires, are typically subjected to high levels of EMI. The input stage of a data acquisition system is usually composed of an instrumentation amplifier which contains an appropriate input circuit, capable of withstanding and filtering high levels of common-mode EMI which is picked up by the wires connecting the sensor(s) to the system in question: this concept is illustrated in Fig. 5.28. EMI is typically present as a common-mode disturbance at the input of the instrumentation amplifier, because the wires running between the actuators and the instrumentation amplifier are routed close to each other and are therefore disturbed in the same way by the surrounding electromagnetic environment. This common-mode EMI can typically attain very high values of several tens of volts: conversely, the sensor output signal is differential (or at least contains a differential component), and usually quite small (some sensing elements have output signals of a few mV) [Cho07]. For these reasons, a large common-mode rejection ratio (CMRR) is necessary to ensure that the measured differential signal is not distorted beyond recognition by the injected common-mode EMI. Component mismatches decrease the CMRR because they introduce a common-mode to differential-mode gain [Ste87]. This case study describes the design of an instrumentation amplifier input circuit presenting a high degree of immunity against common-mode EMI which is injected into its inputs. The proposed input structure cancels the effect of mismatches in the biasing current sources by warping the resulting differential-mode EMI to higher frequencies. Simulations comparing the classic input structure to the proposed input structure are presented: the obtained results illustrate the superior electromagnetic immunity of the proposed instrumentation amplifier input circuit to common-mode EMI, with respect to mismatches between the biasing current sources.

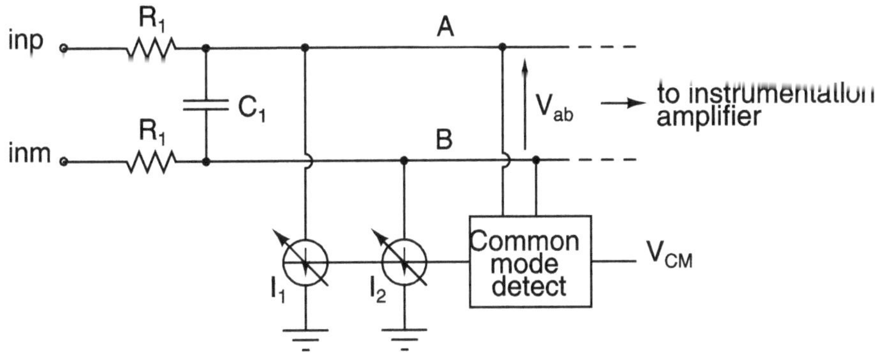


*Figure 5.29.* Ideal instrumentation amplifier input circuit with a high resistance to common-mode EMI injection.

## 3.1 Classic instrumentation amplifier input circuit

A possible instrumentation amplifier input circuit with a high common-mode EMI rejection has been presented in [Com04]. Its working concept is illustrated in Fig. 5.29. In this implementation, the common-mode EMI which is injected in the input terminals *inp* and *inm* is detected by means of a common-mode detect circuit block, and compared to the wanted DC bias voltage (called $V_{CM}$). The common-mode detect circuit drives two matched current sources, $I_1$ and $I_2$, which bias nodes $A$ and $B$ close to $V_{CM}$. Resistors $R_1$ form a high impedance buffer between the IC input pins and the instrumentation amplifier inputs, and protect the latter from high-voltage EMI swings. Owing to inevitable component mismatches, a portion of the common-mode EMI is translated to differential-mode EMI because the common-mode to differential-mode voltage gain ($A_{cd}$) is no longer zero [Ste87]. The CMRR is defined as the ratio between the differential-mode gain ($|A_{dd}|$) and $|A_{cd}|$ [San06]:

$$CMRR\,(\text{dB}) = 20 \cdot \log\left(\frac{|A_{dd}|}{|A_{cd}|}\right) \tag{5.60}$$

Since the input circuit in Fig. 5.29 does not amplify nor filter the wanted differential signals, $|A_{dd}|$ is close to 1. Because the disturbing common-mode EMI signals are much larger than the wanted differential signal, the slightest mismatch converts the large common-mode EMI signals into non-negligible differential signals, which are summed to the wanted input signal. These out-of-band differential EMI signals are partially filtered by the low-pass filter formed by $R_1$ and $C_1$, but this filtering may not be enough to suppress the unwanted differential EMI signals completely. Consequently, the corresponding cut-off frequency formed by $R_1$ and $C_1$ has been placed at 10 kHz. The latter frequency has not been chosen arbitrarily, as it will equally be considered as the

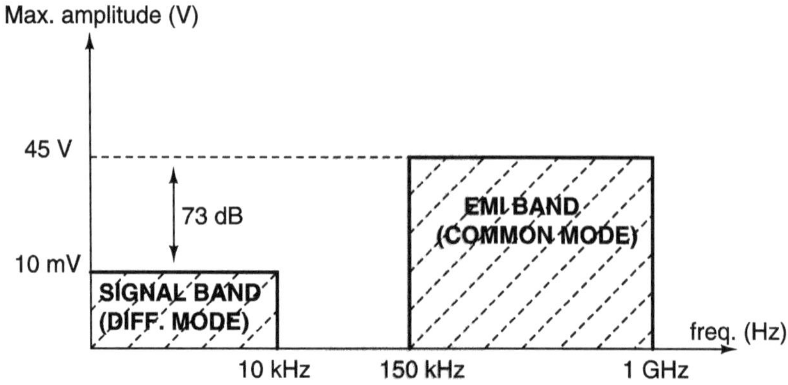


*Figure 5.30.* EMI and signal frequency spectra.

upper limit of the wanted signal band in this present case study. The EMI amplitude has been chosen equal to 45 V, and ranges between 150 kHz and 1 GHz, according to the DPI specification [DPI]. The wanted signal band stretches from 0 to 10 kHz, and the differential amplitude has been chosen equal to 10 mV. This means that the common-mode EMI voltage exceeds the wanted differential signal level by 73 dB. These values are represented schematically in Fig. 5.30.

Owing to this inequity, the slightest mismatch introduces a non-negligible differential EMI component, which mixes with the wanted differential signal and is amplified in the same way further on by the instrumentation amplifier. In addition, the highest (differential) signal frequency and the lowest (common-mode) EMI frequency are approximately separated by a spectral decade (10 kHz and respectively 150 kHz), which poses a severe requirement on the necessary filtering. As is illustrated further on, previously listed constraints have a high impact on the behavior of this input circuit when mismatch is considered. The basic operation of this input circuit is derived first. Refer to the linearized model depicted in Fig. 5.31. In case this circuit is ideally matched, currents $I_1$ and $I_2$ are equal to each other, and expressed as follows:

$$I_1 = I_2 = G_m \cdot \left( \frac{V_A + V_B}{2} - V_{CM} \right) \tag{5.61}$$

where $G_m$ represents the common-mode transconductance of each current source, while $V_{CM}$ is the required DC bias voltage as described previously. In case of ideal matching, the differential voltage $V_{ab}$ contains no EMI contribution, and the common-mode voltages on nodes $A$ and $B$ are identical, and

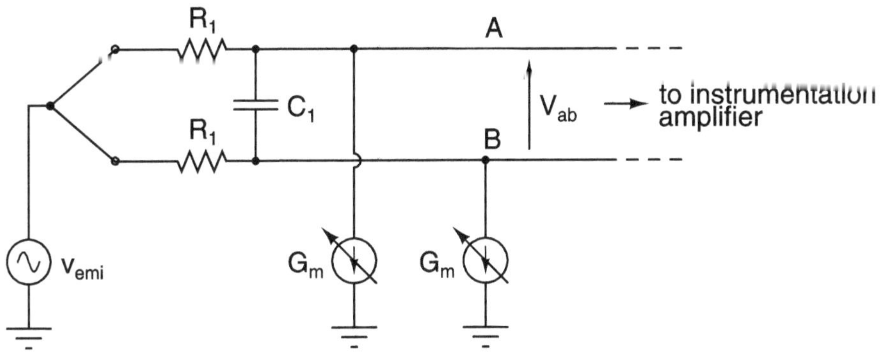


*Figure 5.31.* Ideal instrumentation amplifier input circuit with a high resistance to common-mode EMI injection: linearized equivalent model.

expressed as:

$$V_A = V_B = \frac{G_m \cdot R_1 \cdot V_{CM}}{1 + G_m \cdot R_1} + \frac{v_{emi}}{1 + G_m \cdot R_1} \tag{5.62}$$

Observe that as long as $G_m \cdot R_1 \gg 1$, the DC voltage at nodes $A$ and $B$ is equal to the wanted input DC bias $V_{CM}$, while the common-mode EMI is strongly attenuated by $G_m \cdot R_1$. In addition, the differential voltage $V_{ab}$ is not troubled by the injected EMI, since $V_A = V_B$ in (5.62). This is illustrated in Fig. 5.32: although the EMI common-mode component (45 V at 150 kHz) is much larger than the wanted differential-mode signal (10 mV at 10 kHz), the spectrum of $V_{ab}$ only shows the wanted differential signal spectral line.

The input circuit depicted in Fig. 5.29 works fine when the circuit components are perfectly matched: as derived in (5.62), the common-mode EMI component is strongly attenuated at the inputs of the instrumentation amplifier in that event. This situation changes abruptly when mismatch is considered. Indeed, mismatch transforms a portion of the common-mode disturbance into a differential-mode component owing to a non zero common-mode to differential-mode gain $A_{cd}$. The input circuit shown in Fig. 5.29 has three possible mismatch sources: the common-mode detect circuit block, resistor pair $R_1$ and current sources $I_1$ and $I_2$. These three possible mismatch sources are now examined in detail and their individual contribution to the CMRR is hereby assessed.

- **Mismatch in the common-mode detect circuit block:** Owing to mismatch in the common-mode detect circuit block, a different common-mode voltage ($V_{cm}$) is detected, namely:

$$V_{cm} = \frac{V_A + V_B}{2} + \Delta V_{cm} \tag{5.63}$$

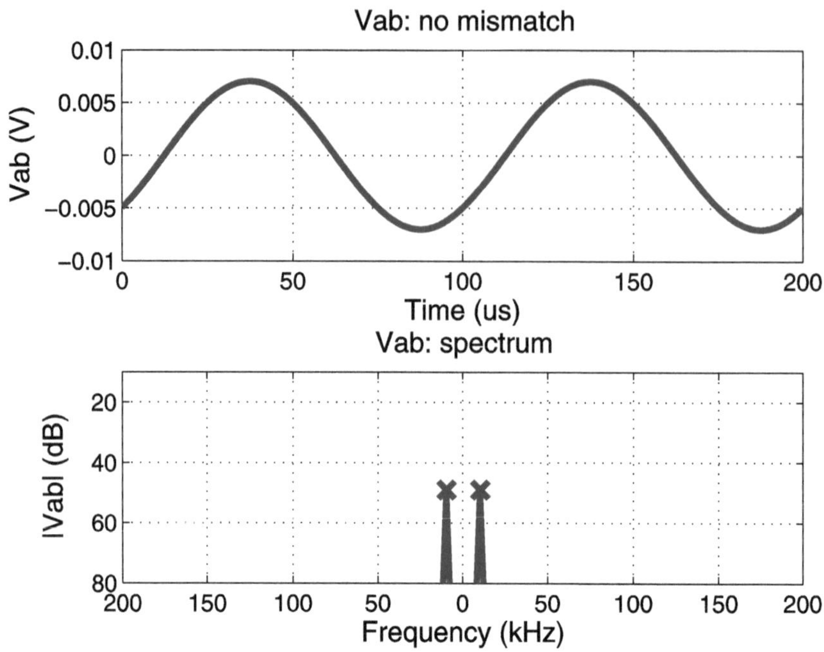


*Figure 5.32.* Transient signal and corresponding spectrum of $V_{ab}$ without mismatch. The common-mode EMI signal is 45 V @ 150 kHz and the wanted differential signal is equal to 10 mV @ 10 kHz. The low-pass filter cut-off frequency is placed at 10 kHz ($R_1 = 100\ \text{k}\Omega$ and $C_1 = 80$ pF).

where $\Delta V_{cm}$ represents the mismatch component. However, as long as both currents $I_1$ and $I_2$ and both $R_1$'s are perfectly matched, the same current flows in or out nodes $A$ and $B$, which means that the voltages on these nodes are equal to each other and that the differential voltage between nodes $A$ and $B$ is zero. Consequently, mismatch in the common-mode detect circuit block does not affect the CMRR.

- **Mismatch between the $R_1$'s:** Assume that there is a mismatch between both $R_1$'s, represented by $\Delta R_1$. Provided that $G_m \cdot R_1 \gg 1$ and that $\Delta R_1 \ll R_1$, the differential voltage between nodes $A$ and $B$ is expressed as:

$$V_{ab} = -\frac{\Delta R_1}{R_1} \cdot V_{CM} + \frac{\frac{\Delta R_1}{R_1} \cdot v_{emi}}{1 + 2 \cdot s \cdot C_1 \cdot R_1} \tag{5.64}$$

Clearly, the mismatch induced differential-mode EMI component ($\frac{\Delta R_1}{R_1} \cdot v_{emi}$) is filtered by low-pass filter $R_1$-$C_1$. Unfortunately, this filtering may not be sufficient to attenuate significantly the low frequency EMI: recall that in a worst case situation, the signal band (10 kHz) and the lowest EMI frequency (150 kHz) are separated by barely a spectral decade, while

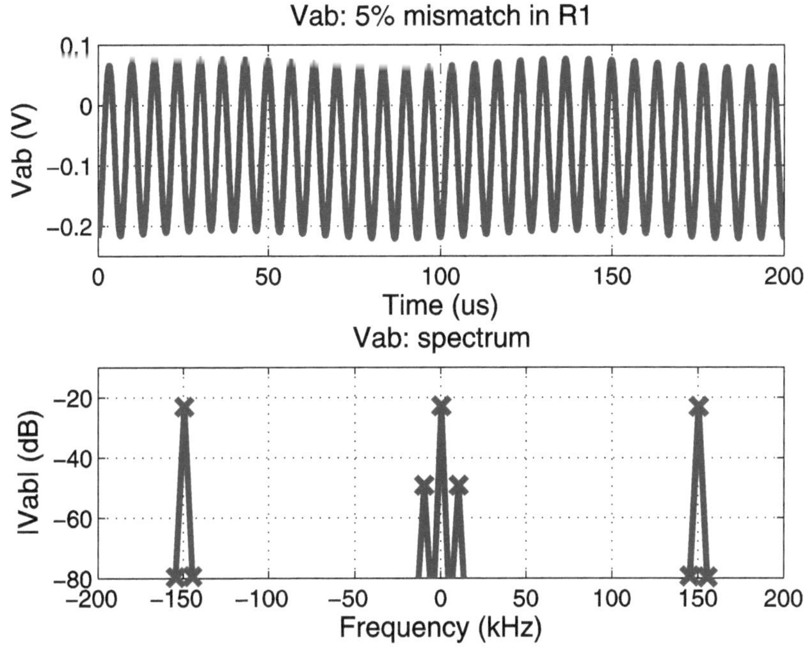


*Figure 5.33.* Transient signal and corresponding spectrum of $V_{ab}$ with a 5% mismatch between the $R_1$'s. The common-mode EMI signal is 45 V @ 150 kHz and the wanted differential signal is equal to 10 mV @ 10 kHz. The low-pass filter cut-off frequency is placed at 10 kHz ($R_1 = 100$ kΩ and $C_1 = 80$ pF). Owing to the mismatch, there is a differential-mode EMI component at 150 kHz, as well as a non-negligible DC component.

the EMI signal amplitude is much larger than the wanted differential signal level (45 V compared to 10 mV, as mentioned previously). In addition, (5.64) shows that the input DC voltage of the instrumentation amplifier is affected as well by this mismatch: however, the appearing DC shift is oblivious to the injected EMI and solely depends on the relative mismatch term $\Delta R_1/R_1$. This DC shift biases the instrumentation amplifier inputs at different DC voltages and is therefore a serious limiting factor. Both effects are clearly visible in Fig. 5.33.

- **Mismatch between the current sources $I_1$ and $I_2$:** Assume that there is a mismatch between the common-mode transconductances $G_m$ of the voltage controlled current sources $I_1$ and $I_2$, represented by $\Delta G_m$. Provided that $G_m \cdot R_1 \gg 1$ and that $\Delta G_m \ll G_m$, the differential voltage between nodes $A$ and $B$ is then expressed as:

$$V_{ab} = -\frac{\Delta G_m}{G_m} \cdot V_{CM} + \frac{\frac{\Delta G_m}{G_m} \cdot v_{emi}}{1 + 2 \cdot s \cdot C_1 \cdot R_1} \tag{5.65}$$

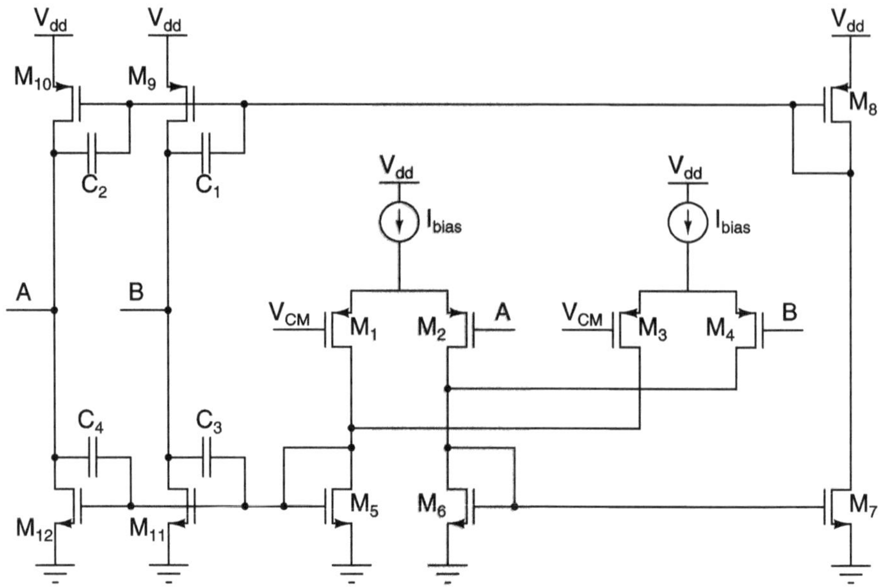


*Figure 5.34.* Common-mode detect circuit and matched current sources.

This expression is very similar to expression (5.64), obtained previously for the mismatch between the $R_1$'s. Likewise, this mismatch generates a differential-mode EMI component as well as an extra DC shift.

Up till now, ideal biasing current sources and an ideal common-mode detect circuit have been used. A practical implementation for the common-mode detect circuit with matched biasing current sources was proposed in [Com04] and is depicted in Fig. 5.34. Theoretically, the bandwidth of the common-mode detect circuit must be higher than the highest EMI frequency in order to remove all the high frequency common-mode EMI components. Knowing that the highest EMI frequency lies at 1 GHz (as described in the DPI specification [DPI]), this may be quite complicated to achieve. This draconian requirement is overcome by shorting the common-mode feedback loop for high EMI frequencies using capacitances $C_1$, $C_2$, $C_3$ and $C_4$. For low EMI frequencies, the common-mode voltages at nodes $A$ and $B$ are small because of the high open loop gain. At higher EMI frequencies, capacitances $C_1$, $C_2$, $C_3$ and $C_4$ take over and short the common-mode signal to ground. Provided that $C_1 = C_2 = C_3 = C_4$, $g_{m1} = g_{m2} = g_{m3} = g_{m4}$, $g_{m5} = g_{m6} = g_{m7} = g_{m8}$ and $g_{m9} = g_{m10} = g_{m11} = g_{m12}$, the closed loop common-mode transconductance $G_m$ is expressed as:

$$G_m = 4 \cdot \frac{g_{m1} \cdot g_{m10}}{g_{m8}} + 2 \cdot s \cdot C_1 \cdot \left(1 - \frac{2 \cdot g_{m1}}{g_{m8}}\right) \tag{5.66}$$

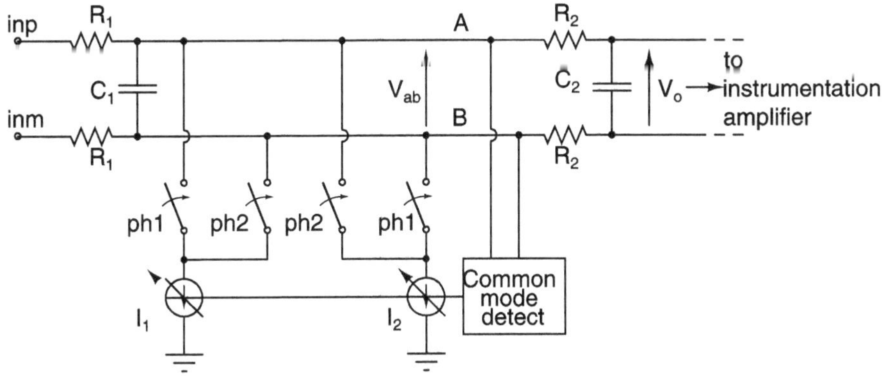


*Figure 5.35.* Instrumentation amplifier input circuit with a high resistance to common-mode EMI injection using switching current sources.

Which illustrates that for high common-mode frequencies, the closed loop common-mode transconductance increases owing to the output capacitors. The stability of this circuit is now evaluated. Under the same assumptions as specified here above, and considering that the circuit open loop output impedance is much higher than $R_1$, the common-mode open loop transfer function of this circuit is found equal to:

$$H_{cm}(j\omega) = \frac{4 \cdot g_{m1} \cdot R_1 \cdot (-g_{m10} + s \cdot C_1)}{g_{m8} \cdot (1 + 2 \cdot s \cdot C_1 \cdot R_1)} \tag{5.67}$$

As can be seen in previous expression, the common-mode open loop voltage transfer function contains a negative pole and a positive zero: however, the latter lies at a much higher frequency and therefore presents no noticeable stability problem, as was verified with simulations.

## 3.2 Input circuit using current sources modulation

As explained previously, mismatch translates a portion of the common-mode EMI to differential-mode EMI. In the previous section, it was derived that the two main mismatch sources are resistor pair $R_1$ and current sources $I_1$ and $I_2$. As will be derived further on, the proposed input circuit topology effectively cancels the mismatch effect attributed to the latter using current modulation. Refer to Fig. 5.35. Current sources $I_1$ and $I_2$ are now switched alternatively between $A$ and $B$ using two non-overlapping clock phases *ph1* and *ph2*. The low frequency differential EMI components are hereby modulated by the switching frequency and warped accordingly to higher frequencies. A small low-pass filter ($R_2$-$C_2$) is added to filter the switching components which are situated at odd multiples of the switching frequency itself. The circuit operation is now derived mathematically. Owing to the mismatch between $I_1$ and $I_2$, $V_{ab}$ can be expressed by the sum of three terms: firstly the wanted differential signal ($A \cdot \sin(\omega t)$), secondly the DC shift which purely depends on the relative mismatch itself (represented by $V_{OFF}$, refer to the DC term in expression (5.65)) and thirdly the EMI differential-mode component which is affected by the EMI amplitude as well as by the relative mismatch between $I_1$ and $I_2$ (represented by $B \cdot \sin(\omega_{emi} t)$, refer to the AC term in expression (5.65)). This yields:

$$V_{ab} = A \cdot \sin(\omega t) + V_{OFF} + B \cdot \sin(\omega_{emi} t) \tag{5.68}$$

Assume that both current sources are periodically switching with period $T_c$. $V_{ab}$ can then be expanded as:

$$\begin{aligned} V_{ab} &= A \cdot \sin(\omega t) \\ &\quad + \begin{cases} V_{OFF} + B \cdot \sin(\omega_{emi} t) & 0 < t < \frac{T_c}{2} \\ -V_{OFF} - B \cdot \sin(\omega_{emi} t) & \frac{T_c}{2} < t < T_c \end{cases} \end{aligned} \tag{5.69}$$

As expressed in (5.69), the (mismatch induced) second and the third terms of $V_{ab}$ are modulated with a square wave with 1 V amplitude. This square wave can be represented by its Fourier series in the following way [Arf95]:

$$V_{square}(t) = \frac{4}{\pi} \sum_{n=1,3,5,\ldots}^{\infty} \frac{\sin(n\omega_c t)}{n} \tag{5.70}$$

where $\omega_c = \frac{2\pi}{T_c}$. Combining (5.69) and (5.70), $V_{ab}$ is expressed as follows:

$$\begin{aligned} V_{ab} &= A \cdot \sin(\omega t) + (V_{OFF} + B \cdot \sin(\omega_{emi} t)) \cdot V_{square}(t) \\ &= A \cdot \sin(\omega t) + \frac{4}{\pi} \cdot (V_{OFF} + B \cdot \sin(\omega_{emi} t)) \cdot \sin(\omega_c t) \\ &\quad + \frac{4}{3\pi} \cdot (V_{OFF} + B \cdot \sin(\omega_{emi} t)) \cdot \sin(3\omega_c t) + \cdots \end{aligned} \tag{5.71}$$

Not surprisingly, previous equation shows that the original EMI signal is amplitude modulated by odd multiples of switching frequency $\omega_c$. Indeed, expanding (5.71) yields:

$$\begin{aligned} V_{ab} &= A \cdot \sin(\omega t) + \frac{4 \cdot V_{OFF}}{\pi} \sin \omega_c t \\ &\quad + \frac{2 \cdot B}{\pi} (\cos((\omega_c - \omega_{emi})t) - \cos((\omega_c + \omega_{emi})t)) + \cdots \end{aligned} \tag{5.72}$$

Observe that a low frequency EMI signal is warped to higher frequencies where it is more attenuated owing to low-pass filters $R_1$-$C_1$ and $R_2$-$C_2$. The cut-off frequency of low-pass filter $R_2$-$C_2$ can be chosen much higher than the cut-off frequency of low-pass filter $R_1$-$C_1$, since its primary function is to filter

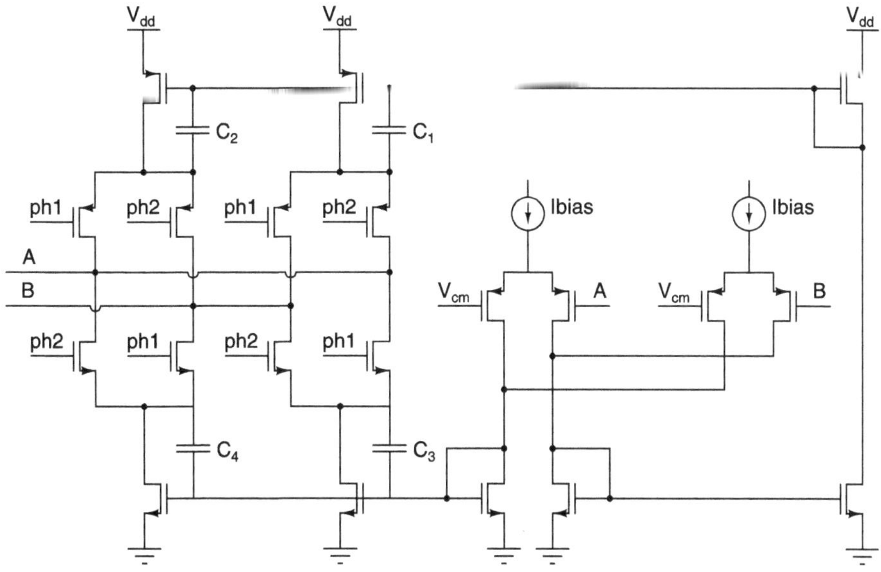


*Figure 5.36.* Common-mode detect circuit and matched current sources, using switching to attenuate mismatch effects in the biasing current sources.

the odd multiples of the high frequency switching component. As can be seen in (5.71) and (5.72), the magnitudes of the switching components are proportional to the offset voltage $V_{OFF}$, which is in turn generated by the relative mismatch term. In addition, as will be clarified in the next paragraph, low-pass filter $R_2$-$C_2$ does not introduce additional mismatch components. Conversely, high frequency EMI components are aliased back into the signal band: however, they are strongly attenuated by $R_1$-$C_1$ and are therefore negligible, as will be illustrated with a simulation example further on. Mismatch in the common-mode detect circuit does not introduce DC shift nor generate differential EMI components. This leaves three possible sources of mismatch: the switched current sources $I_1$ and $I_2$, resistor pair $R_2$ and resistor pair $R_1$. These three potential mismatch sources are now evaluated in detail.

- **Mismatch between the switched current sources $I_1$ and $I_2$:** As derived in the previous section, switching warps the differential-mode EMI as well as the DC shift which are both induced by mismatch between $I_1$ and $I_2$ to higher frequencies, where they are attenuated by low-pass filters $R_1$-$C_1$ and $R_2$-$C_2$. The higher the switching frequency, the more effective the filtering: however, a high switching frequency complicates the design of the switches and increases amongst others the risk of clock feedthrough, which generates other non-idealities in the circuit's behavior [San06].

*Table 5.3.* Design specifications.

| | *Classic input circuit* | *Input circuit with switching current sources* |
|---|---|---|
| $R_1$ | 100 kΩ | 100 kΩ |
| $C_1$ | 80 pF | 80 pF |
| $R_2$ | – | 100 kΩ |
| $C_2$ | – | 8 pF |
| $f_{switching}$ | – | 1 MHz |

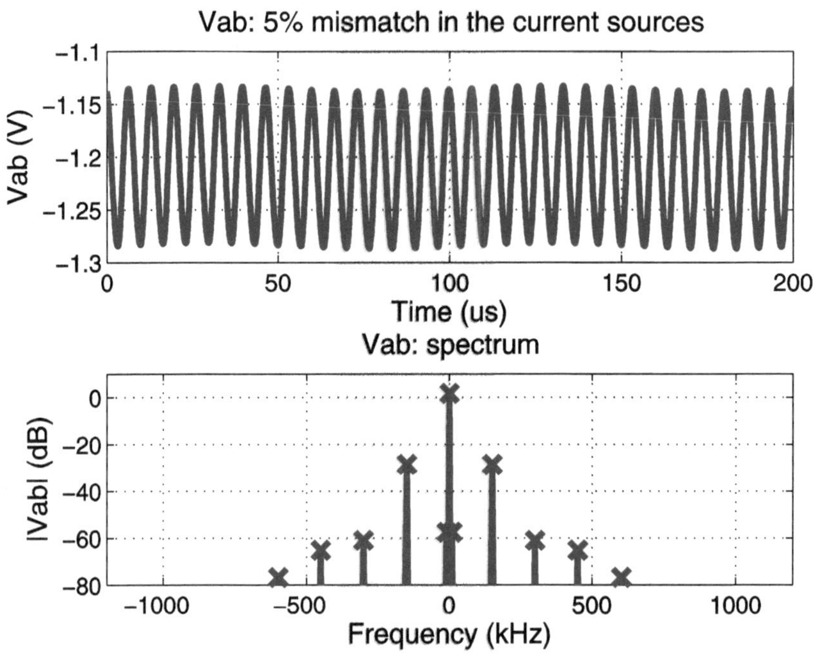


*Figure 5.37.* Transient signal and corresponding spectrum of $V_{ab}$ with 5% mismatch between the current sources using the classic circuit. The common-mode EMI signal is 45 V @ 150 kHz and the wanted differential signal is equal to 10 mV @ 10 kHz. The magnitude of the wanted differential signal (10 kHz) equals −57.5 dB, while the fundamental EMI frequency (150 kHz) equals −28.5 dB. The DC component is very large (1.7 dB).

- **Mismatch between the $R_2$'s:** Provided that the input impedance of the instrumentation amplifier is high, there is no common-mode to differential-mode gain. Mismatch between resistors $R_2$ does therefore not contribute to the CMRR.

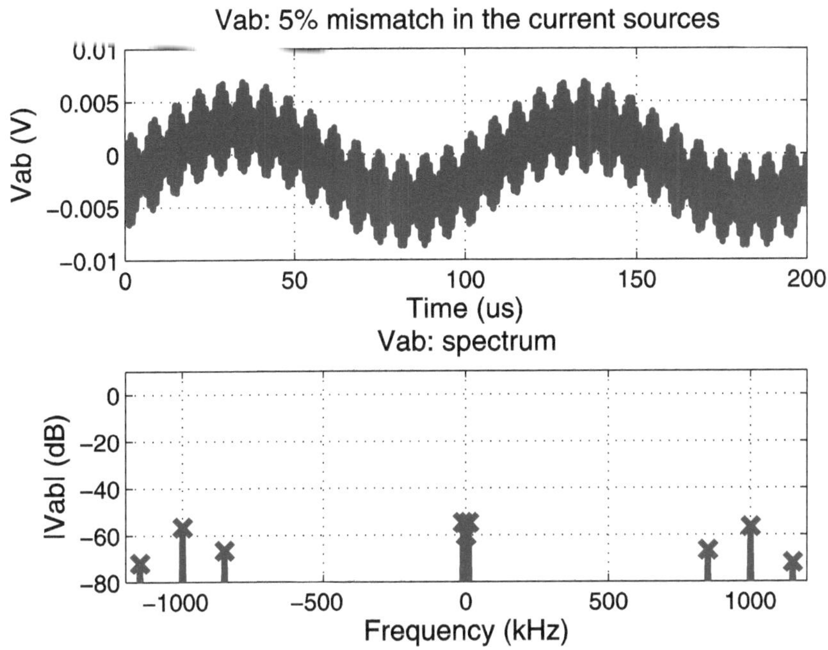


*Figure 5.38.* Transient signal and corresponding spectrum of $V_{ab}$ with 5% mismatch between the current sources using the switching topology. The common-mode EMI signal is 45 V @ 150 kHz and the wanted differential signal is equal to 10 mV @ 10 kHz. The magnitude of the wanted differential signal (10 kHz) equals −55 dB, while the fundamental EMI frequency (150 kHz) has been shifted to 850 kHz (−66.6 dB) and 1.15 MHz (−71.8 dB). The DC component has been fully warped to the switching frequency of 1 MHz (−56.6 dB).

- **Mismatch between the $R_1$'s:** Differential EMI components as well as the DC shift caused by mismatches between the $R_1$'s are not warped to higher frequencies by the switching current sources because the common-mode to differential-mode translation occurs anteriorly to the switching. Therefore, extra care must be insured to match $R_1$ accordingly. This is confirmed by simulations.

The current sources in Fig. 5.34 are modified so as to switch alternatively between $A$ and $B$. Clock phases *ph1* and *ph2* are complementary and non-overlapping. A practical transistor implementation is shown in Fig. 5.36.

## 3.3 Simulations

The classic instrumentation amplifier input circuit (Fig. 5.29) has been compared to the proposed input circuit which switches the biasing current sources (Fig. 5.35). Both circuits using the corresponding common-mode detect and matched current sources circuit blocks (depicted in respectively Figs. 5.34

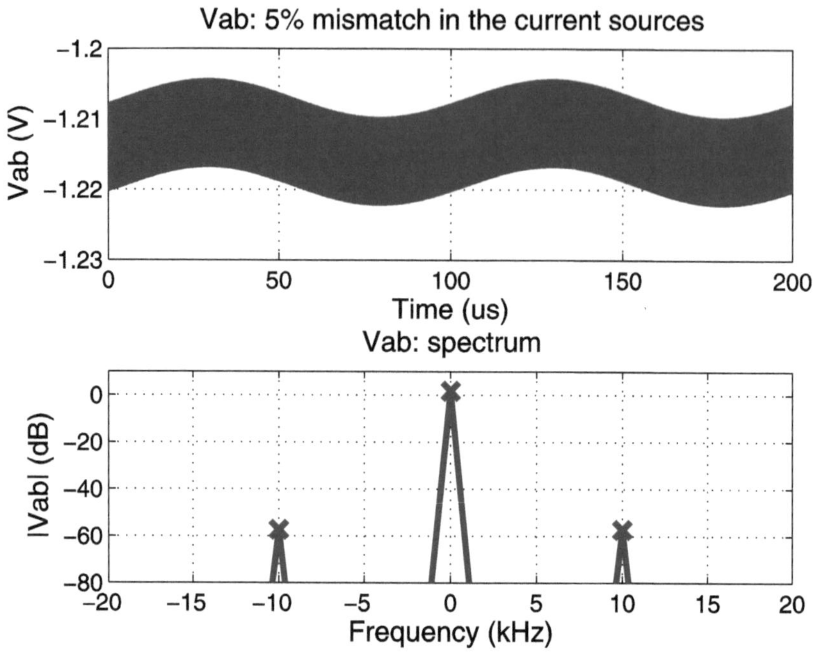


*Figure 5.39.* Transient signal and corresponding spectrum of $V_{ab}$ with 5% mismatch between the current sources using the classic circuit. The common-mode EMI signal is 45 V @ 2 MHz and the wanted differential signal is equal to 10 mV @ 10 kHz. The magnitude of the wanted differential signal (10 kHz) equals −57.5 dB, while the fundamental EMI frequency (2 MHz) is heavily attenuated, and equals −50.8 dB (not shown on the plot). However, the DC component is again very large (1.7 dB).

and 5.36) have been simulated a standard 0.35 μm CMOS technology. The design specifications are enumerated in Table 5.3.

Low-pass filter $R_1$-$C_1$ is designed for a cut-off frequency of 10 kHz, since this frequency constitutes the highest frequency in the considered signal band. The switching frequency has been chosen equal to 1 MHz, which allows sufficient EMI suppression as is illustrated shortly. Low-pass filter $R_2$-$C_2$ has been designed for a cut-off frequency of 100 kHz, since its primary function is to attenuate the odd multiples of the switching frequency.

Figures 5.37 and 5.38 depict the transient signal and the corresponding frequency spectrum when a common-mode EMI of 45 V amplitude at a frequency of 150 kHz is injected respectively in the input pins of the classic and the proposed input circuit when there is a 5% mismatch between the matched current sources. Observe in Fig. 5.38 that owing to the switching, the EMI low frequency component of 150 kHz is warped to 1.15 MHz and 850 kHz, while the DC shift appears on the 1 MHz spectral line. These components are in turn strongly attenuated by low-pass filters $R_1$-$C_1$ and $R_2$-$C_2$. A second simula-

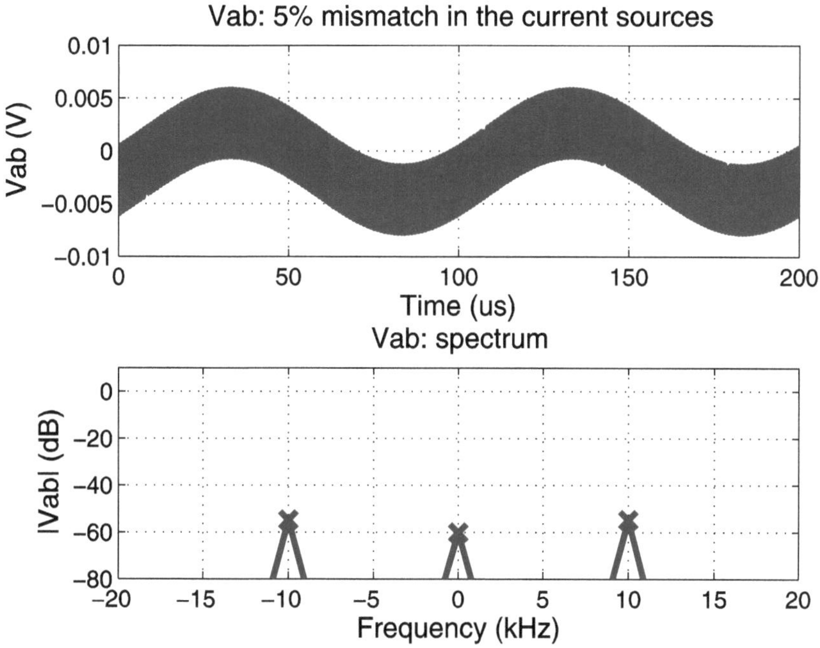


*Figure 5.40.* Transient signal and corresponding spectrum of $V_{ab}$ with 5% mismatch between the current sources using the switching topology. The common-mode EMI signal is 45 V @ 2 MHz and the wanted differential signal is equal to 10 mV @ 10 kHz. The DC component has been warped to the switching frequency (1 MHz), and equals $-56.2$ dB. The EMI component of 2 MHz which is folded back to DC is quite small ($-60.2$ dB).

tion illustrates the effect of EMI which is folded back into the signal band: the EMI frequency has been chosen equal to 2 MHz, with the same amplitude (45 V). Figures 5.39 and 5.40 compare the transient signal and the corresponding output spectrum for respectively the classic and the proposed input circuit. Observe that the DC shift is strongly attenuated in the proposed input circuit, as can be appreciated in Fig. 5.40. The small DC component of $-60$ dB is the 2 MHz EMI signal which is folded into baseband.

# Chapter 6

# EMI Resisting Bandgap References and Low Dropout Voltage Regulators

*The ban on the use of mobile phones by passengers on planes is set to continue. New tests by the Civil Aviation Authority confirmed that phones are still a threat to aircraft. The latest study found that the use of mobile telephones can adversely affect navigation and communication functions, producing significant errors on instrument displays and background noise on audio outputs. (...) The research backs up reports from pilots, who have stated that interference from mobiles has caused: false notification of unsafe conditions – for example, incorrect baggage compartment smoke alarm warnings; malfunction of aircraft systems; interrupted communications due to noise in the flight crew headphones; distraction of crews from their normal duties due to increased work levels and the possibility of having to invoke emergency drills. (...)*

—Quoted from [Arm07]

## 1 Introduction

Analog integrated circuits are not solely disturbed at their output and input terminals, as discussed respectively in Chaps. 4 and 5: they are equally very susceptible to EMI which is injected into their power supply pins. These disturbances are typically generated by fast switching transients, produced amongst others by digital circuits, RF subcircuits and DC-DC converters, drawing short and steep current spikes from the power supply [Max01, Gup05]. This resulting conductive noise deteriorates the operation of sensitive analog circuits like synthesizers and voltage controlled oscillators (VCO's), and manifests itself as jitter on their outputs, impacting critical system requirements. Even worse, and hardly surprising in view of the previous case studies, EMI injected in the supply can trigger adverse DC shift phenomena in precarious nonlinear circuit nodes, which debias the circuit by altering its correct DC bias points, as explained in Chap. 3.

J.-M. Redouté, M. Steyaert, *EMC of Analog Integrated Circuits,*
Analog Circuits and Signal Processing,
DOI 10.1007/978-90-481-3230-0_6, © Springer Science+Business Media B.V. 2010

Linear voltage regulators are entrusted with the important task of shielding sensitive circuit blocks from the noisy supply lines, by converting a given DC supply input voltage into a specific and stable DC output voltage, as well as maintaining it over a wide range of load conditions [Max01, Gup05]. For instance, switching regulators are used in mobile phones to boost the supply voltages to higher values, and subsequently, low dropout voltage (LDO) regulators are cascaded in series in order to suppress the inherent noise associated to the switching devices [Rin98]. Unfortunately, voltage regulators are in turn highly susceptible to EMI, since their input terminal constitutes the noisy power supply line itself. This predisposition is best explained by considering their intrinsic structure. A typical voltage regulator block diagram is depicted in Fig. 6.1: it consists of the following circuit parts [Spe77].

- **A reference element,** providing a stable and known reference voltage level ($V_{ref}$). Nowadays, high precision on-chip bandgap reference circuits are typically used to generate the required reference voltage: these derive their name from the energy bandgap voltage, which is equal to 1.206 V in silicon at zero Kelvin [Wid71]. They can be realized to generate very precise voltages, quasi-independently of the ambient temperature, as illustrated in [Bec08].

- **A sample or feedback circuit,** sensing a fraction of the output voltage. In most cases, a simple resistive divider may be used for that purpose. Systematic offset does not impact the resistive division to a large extent, because the generated feedback voltage is determined by the resistor ratio, instead of absolute resistor values. This resistor ratio can generate an error which is smaller than 0.1%, provided that proper layout techniques are used [San06].

- **An operational amplifier,** comparing the sampled voltage to the reference voltage, and generating a corresponding error signal. Any offset voltage is viewed by the opamp as an error signal, causing the output to respond accordingly. In case a high precision is required, low offset opamp design techniques can be used to decrease the opamp offset to very small (sub-microvolt) values [Enz96].

- **A pass device,** providing the translation of the input signal to the desired output voltage level over varying load conditions. This pass device behaves as a resistor in a linear voltage regulator, or as a switch in a switching regulator.

Even though the regulation methods vary amongst the different voltage regulator types (e.g. shunt, series, switching, ...), these four basic functions are present in all basic regulator circuits using feedback [Spe77]. Voltage regulators have been designed without feedback as well, but since these do not offer

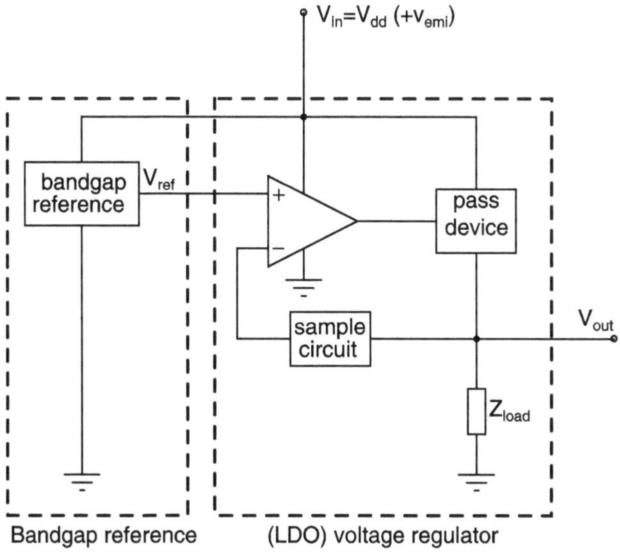


*Figure 6.1.* Basic voltage regulator block diagram.

a high degree of accuracy, they are not considered in the remainder of this chapter [Man83]. Clearly, when EMI is injected into the positive supply rail (represented by $v_{emi}$ in Fig. 6.1), it interferes with the load in three distinct ways:

- First of all, EMI disturbs the bandgap reference circuit: this is particularly harmful since in that event, the opamp imposes a noisy or altered output voltage $V_{out}$ across the load. This directly impacts the voltage regulator precision.

- Secondly, the EMI source influences the correct operation of the opamp because of the finite PSRR of the latter [Ste90, Ste91, Gie93]. Owing to this, the EMI couples to the control node of the pass device, and consequently interferes with its operation. Cascode devices can be used to increase the power supply rejection (PSRR) [Rib84]: however, they reduce the dynamic range of the opamp because of the increased voltage drop across them, making this stage more prone to EMI induced DC shift [1]. Furthermore, at very high EMI frequencies, cascode transistors don't shield EMI significantly any longer, since their parasitic capacitances become predominant

[1] In particular, transistors are forced more rapidly out of their operating region owing to the smaller voltage headroom.

[Ste90]. In addition, the PSRR is in practice always limited by mismatches [Gie93].

- Thirdly, the EMI source couples across the pass device to the load owing to the presence of parasitic capacitances of the latter (e.g. through the parasitic drain-gate and drain-bulk capacitances of a NMOS pass device transistor). Here too, cascoded devices may be used in order to shield the load from the EMI source. But aside from increasing the total voltage drop, they are only effective at low EMI frequencies [Ste90].

Bandgap reference voltage circuits are not exclusively used in voltage regulators: they are frequently used as standalone circuits as well, when a pure reference voltage is needed which is not required to drive a low impedance (for instance to provide a stable reference voltage in AD and DA converters [Cha08]). To this end, bandgap reference circuits are considered separately in this chapter. Two case studies are described further on:

- **Case study 1: EMI resisting bandgap voltage references.** The first case study evaluates the effect of conducted EMI which is injected in the supply terminal of a classic Kuijk bandgap reference voltage circuit. A first improvement is made by modifying the pass device type and by using the compensation capacitor of the opamp to keep the gate-source voltage of the former at a constant value, hereby cancelling the impact of low frequency EMI on the drain current. However, this structure is still susceptible to high frequency EMI disturbances because of the nonlinear output resistances of the input transistors. The second circuit improvement introduces an active load, which uses a stable positive feedback loop to shield the input differential pair stage from EMI while keeping the average drain currents constant. Measurements of a test-chip corroborate and confirm the theoretical deductions.

- **Case study 2: EMI resisting LDO voltage regulators.** The second case study focuses on designing a low dropout voltage regulator with a high immunity against EMI. Low dropout voltage regulators are particularly praised since they require a small voltage drop across the pass device, enabling the regulator to work with small and ever decreasing supply voltages, while minimizing at the same time the dissipated power across the pass device, which improves their efficiency [Rin98]. The EMI coupling parasitically across the pass device as well as through the OTA owing to its finite PSRR, is suppressed by modifying the pass transistor type and by shorting its gate-source voltage in a similar way as in the first case study. However, owing to parasitic capacitances, unwanted EMI coupling paths from the power supply rail to the output terminal still subsist. A theoretical analysis explores which parameters are crucial in order to reasonably reduce the EMI coupling. Circuit simulations confirm the mathematical analyses.

## 2 Case study 1: CMOS bandgap voltage references with a high immunity to EMI

The theoretical principles of bandgap voltage references, compensating the inverse dependence of temperature on the voltage across a p-n junction, have been developed in the sixties, while the first integrable circuit solutions have emerged in the early seventies [Hil64, Wid71, Kui73]. Bandgap voltage reference circuits are frequently used in modern VLSI applications: in practice, the resolution of modern integrated data acquisition circuits, like AD and DA converters, is limited by the precision of the generated reference voltage [Pre06]. For this reason, they are usually required to function appropriately within tight and well specified tolerances. In order to comply to the strict requirements they are subjected to, bandgap voltage references use temperature compensation and curvature correction techniques to achieve very high precision reference voltages for a broad range of temperature conditions: techniques increasing their precision over a high temperature range have been successfully applied for quite some time now [Son83]. In today's system-on-chip (SOC) applications, bandgap references are very often integrated on the same chip, meaning that a high level of EMI robustness is mandatory [Fio04].

Paradoxically, although much attention is paid to many precision enhancement aspects of bandgap reference circuits, little or no EMI influence is usually considered during their design [Pre06]: paradoxically, EMI may impact and modify the reference voltage orders of magnitude more than e.g. a varying temperature range. Nowadays, with effective temperature coefficients lying around a few $\mathrm{ppm}/^{\circ}\mathrm{C}$ (as in [Bec08]), a strong EMI injection in the power supply easily perturbs these very high precision circuits. Bandgap voltage reference circuits achieving an increased PSRR have been designed using an extra pass device between the supply voltage and the bandgap circuit itself, regulated by a fast feedback loop [Tha95]: although highly effective for lower disturbing frequencies, the high PSRR drops at higher EMI frequencies because the parasitic capacitances become predominant [Ste90]. Additionally, since CMOS device dimensions and corresponding supply voltages are continuously shrinking, fractional bandgap circuits are used to generate sub-1V reference voltages [Per07]: these smaller voltages are in turn more easily disturbed by EMI injected on the power supply, since less shielding devices can be foreseen internally between the reference node and the noisy supply voltage line in question, owing to the reduced voltage headroom.

For the above reasons, present case study describes and introduces circuit solutions improving the global electromagnetic immunity of Kuijk bandgap reference topologies. In order to simplify the already quite complex EMC analysis, the focus is fully placed on obtaining a high EMI suppression: circuit solutions contributing to an increased precision (like curvature correction) have therefore been omitted from the considered designs, since they are extensively documented in the specialized literature [Bec08, Son83]. Finally, the principles and concepts which are introduced in this case study can be extrapolated to a variety of non-Kuijk bandgap voltage reference topologies.

## 2.1 EMI injection in a Kuijk bandgap reference (NPD)

Amongst the various existing bandgap voltage reference architectures, one of the most prevalent ones is undoubtingly the Kuijk bandgap reference circuit, which uses the difference in bias currents through two p-n junctions in combination with an operational amplifier in order to generate the necessary bandgap voltage [Kui73]. In classical bandgap circuits, the reference voltage is produced by summing the base-emitter voltage of a bipolar transistor to the difference between the base-emitter voltages of two transistors with a different emitter current [Wid71]. Since the base-emitter voltage difference has a positive temperature coefficient while the base-emitter voltage has a negative temperature coefficient, the sum of both can be ideally made independent of the temperature as long as the quadrature term is disregarded: in the literature, such behavior is respectively categorized as PTAT (proportional to absolute temperature), and inversely PTAT [San06]. The sum of these two voltages equals the silicon bandgap voltage at zero Kelvin, which is 1.206 V. What differs from circuit to circuit is how the sum between the PTAT and the inverse PTAT term is realized.

A prevailing circuit solution is to use the topology introduced by Kuijk, in 1973 [Kui73]: Figure 6.2 depicts a classic Kuijk bandgap reference voltage circuit. The principle behind this circuit is as elegant as it is straightforward. The generated output voltage $V_{ref}$ can be expressed as the sum of the voltage drops across $R_1$ ($V_{R_1}$), $R_3$ ($V_{R_3}$) and the base-emitter voltage of $Q_1$ ($V_{be1}$). This gives:

$$V_{ref} = V_{R_1} + V_{R_3} + V_{be1} \tag{6.1}$$

As long as the opamp gain is very high, both opamp inputs are virtually shorted, and $V_{R_3}$ is expressed using the p-n junction equation [Gra01]:

$$V_{R_3} = V_{be2} - V_{be1} = \frac{k \cdot T}{q} \cdot \ln\left(\frac{I_2}{I_1} \cdot \frac{I_{S1}}{I_{S2}}\right) = \frac{k \cdot T}{q} \cdot \ln\left(\frac{R_1}{R_2} \cdot \frac{I_{S1}}{I_{S2}}\right) \tag{6.2}$$

Where $I_{S1}$ and $I_{S2}$ represent the saturation currents of $Q_1$ and $Q_2$. Substituting (6.2) in (6.1), and observing that the current through $R_1$ is equal to the current through $R_3$, yields [Kui73, Gra01]:

$$V_{ref} = V_{be1} + \frac{k \cdot T}{q} \cdot \left(1 + \frac{R_1}{R_3}\right) \cdot \ln\left(\frac{R_1}{R_2} \cdot \frac{I_{S1}}{I_{S2}}\right) \tag{6.3}$$

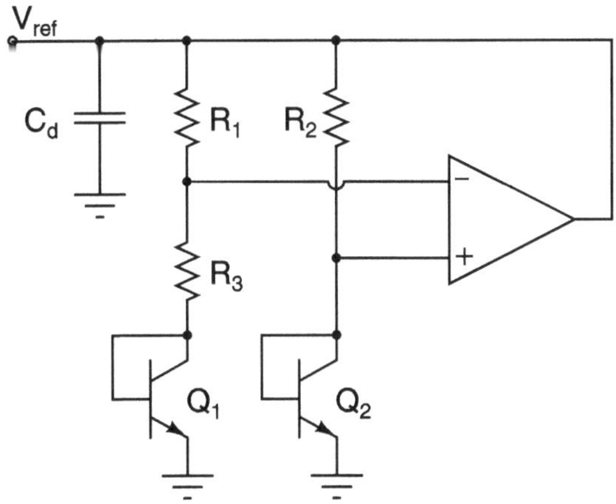


*Figure 6.2.* Classic Kuijk bandgap voltage reference circuit.

As can be seen in previous equation, the first term is the base-emitter voltage of transistor $Q_1$ which is inversely PTAT, while the second term is PTAT: the sum of both can therefore be trimmed to cancel out the inverse temperature dependence of $V_{be1}$.

In practice, the opamp is non ideal, and has a finite voltage gain, with a limited driving capability (meaning a nonzero output impedance). In order to keep the subsequent EMI analysis compact, it has been considered that the opamp is a basic one stage OTA, biased actively by means of a current mirror, driving a pass device transistor. The latter is very often realized as a NMOS transistor connected in a source follower configuration: this circuit is depicted in Fig. 6.3a, and has been extensively mentioned and studied in the specialized literature [San06, Gra01, Raz01]. This classic topology is referred to as the NMOS pass device (NPD) bandgap circuit in the remainder of this chapter. Evidently, when EMI is injected in the positive power supply terminal of this circuit (represented by $v_{emi}$ in Fig. 6.3a), the generated reference voltage $V_{ref}$ is polluted by undesirable EMI components. The resulting high frequency ripple on $V_{ref}$ is particularly harmful for delicate circuits requiring a very precise and constant reference voltage. Even worse, the nonlinear distortion of this EMI signal may alter the DC value of the reference voltage $V_{ref}$: this effect is much worse, since in the latter case, $V_{ref}$ can not be used as an accurate reference any longer. Furthermore, once DC shift has taken place, no ulterior filtering will restore the original voltage. It is therefore of paramount importance to minimize and contain the DC shift as much as possible, while alleviating the ripple appearing on $V_{ref}$ using linear filtering. Observe in Fig. 6.3a that no decoupling capacitor filtering the supply voltage is present. Although such a capacitor is highly effective to filter out high frequency EMI, present case

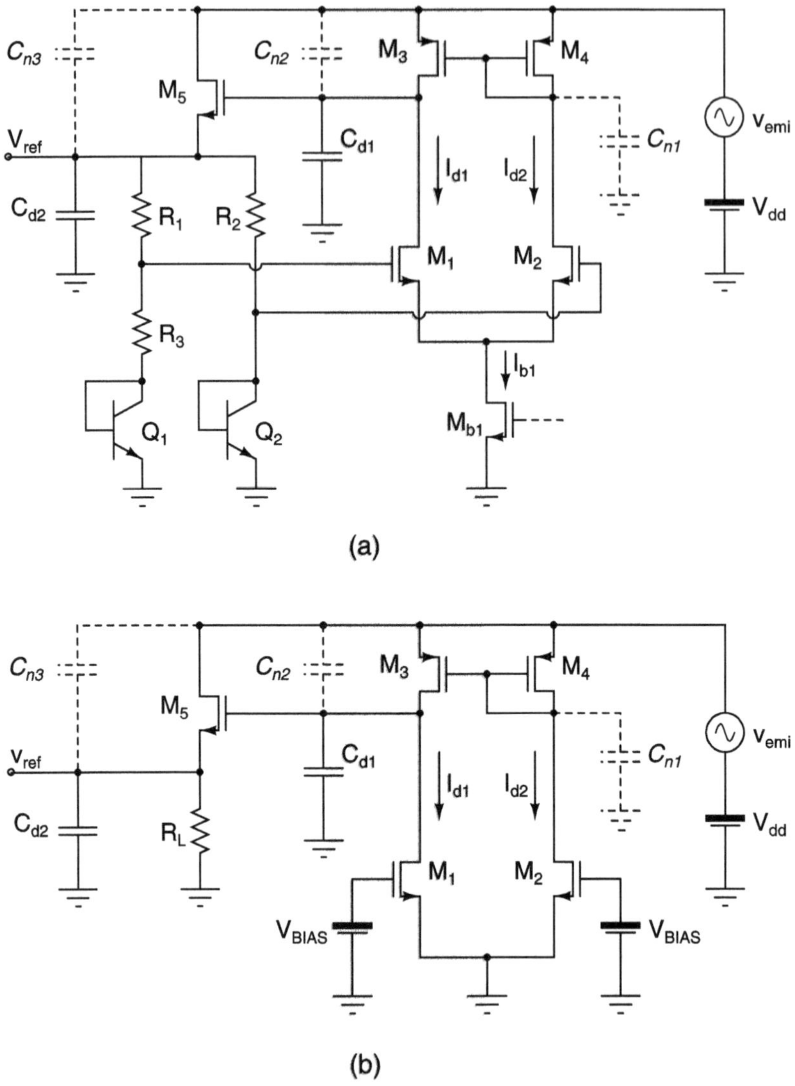


*Figure 6.3.* Kuijk bandgap voltage reference circuit (a) Basic schematic with a NMOS pass device (NPD). (b) Small signal analysis simplified schematic of the NPD bandgap circuit.

study is focused on examining and solving potential EMI threats for the full EMI frequency range (150 kHz – 1 GHz, as specified in the DPI specification [DPI]). For this reason, the presence of a capacitor decoupling the supply lines has been left out in present case study as well as in the designed test-IC. However, in a practical circuit, it is very advisable to decouple the power supply, using internal and external capacitors. Since decoupling capacitors attenuate high frequency EMI disturbances, the focus of subsequent analyses concentrate on attenuating the low frequency EMI signals using circuit techniques.

Following the approach pursued in the previous chapters, small and large signal analyses are derived separately, in order to detect the appearing EMI problems: as customary, the small signal behavior is discussed first.

### 2.1.1 Small signal analysis

As long as the injected EMI level is sufficiently small, the open loop transfer function from the EMI source which is superposed on the supply voltage ($v_{emi}$) to the reference voltage output node ($v_{ref}$) may be calculated using a small signal approach. Five basic assumptions were made in order to simplify the calculation of this open loop transfer function.

- Firstly, the feedback loop was interrupted at the gates of the input pair and replaced by a DC voltage source $V_{BIAS}$, as indicated in Fig. 6.3b. Observe that the calculated attenuation for low EMI frequencies differs from the one which is obtained when the complete closed loop circuit is considered. This difference is, however, not particularly important since it manifests itself solely at low EMI frequencies, lying below the OTA gain bandwidth product (GBW) frequency.
- Secondly, the bandgap cell (consisting of resistors $R_1$, $R_2$ and $R_3$, and transistors $Q_1$ and $Q_2$) is modeled by an equivalent resistor $R_L$. This is a correct approach as long as the diode connected bipolar transistors are linearized in their operating point.
- Thirdly, the dominant parasitic capacitances in this circuit are represented by $C_{n1}$, $C_{n2}$ and $C_{n3}$ (dashed).
- Fourthly, the following analysis considers that the output resistances ($r_o$) and transconductances ($g_m$) of all MOS transistors used in this schematic are equal to each other.
- Fifthly, the output impedance of the current source transistor ($M_{b1}$) has been disregarded, in order to eliminate an extra circuit node. This has only a minor impact on the calculations, since the signal transfer from the EMI source $v_{emi}$ to the drain of $M_1$ is largely dependent on diode connected transistor $M_4$ at low EMI frequencies, and on $C_{n2}$ at high EMI frequencies.

These simplifications are represented schematically in Fig. 6.3b. The resulting simplified open loop transfer function representing the coupling from the positive power supply to the reference voltage $v_{ref}$, is of the second order, and contains two poles and two zeros:

$$A_{DD1}(s) = \frac{v_{ref}(s)}{v_{emi}(s)} = \frac{R_L \cdot g_m}{1 + R_L \cdot g_m} \cdot \frac{(\frac{s}{\omega_{z1}} + 1) \cdot (\frac{s}{\omega_{z2}} + 1)}{(\frac{s}{\omega_{p1}} + 1) \cdot (\frac{s}{\omega_{p2}} + 1)} \tag{6.4}$$

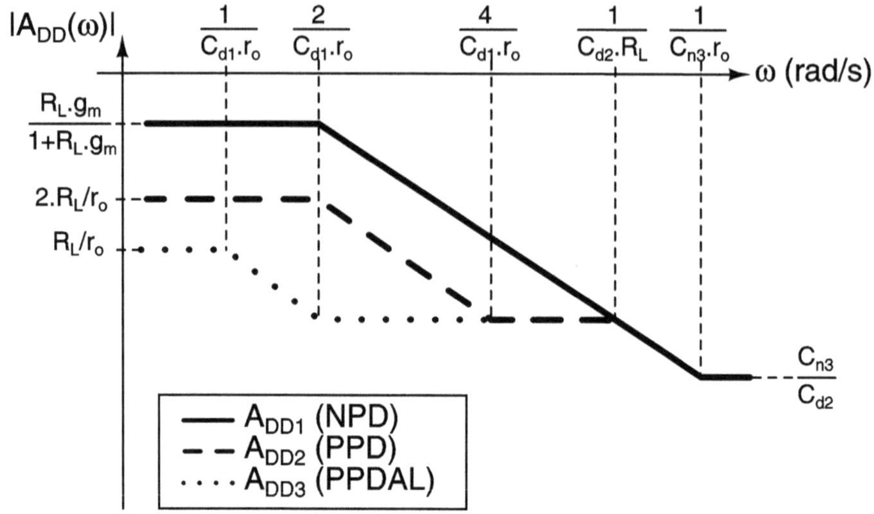


*Figure 6.4.* Comparative Bode plots of the open loop transfer functions associated to the NPD (solid), PPD (dashed) and PPDAL (dotted) bandgap reference circuits.

The pole and zero frequencies are respectively expressed as follows:

$$\begin{cases} \omega_{p1} = \frac{2}{C_{d1} \cdot r_o}, & \omega_{p2} = \frac{g_m \cdot R_L + 1}{C_{d2} \cdot R_L} \\ \omega_{z1} = \frac{1}{C_{n3} \cdot r_o}, & \omega_{z2} = \frac{2 \cdot g_m}{C_{d1}} \end{cases} \quad (6.5)$$

Observe that $\omega_{p2}$ and $\omega_{z2}$ lie close to each other. Therefore, in view of this simplified analysis, it will be considered that they cancel each other out. Figure 6.4 depicts the resulting theoretical Bode plot of this open loop transfer function. Observe that the dominant pole of $A_{DD1}(s)$ ($\omega_{p1}$) coincides with the dominant pole of the opamp and that the ratio between $C_{n3}$ and $C_{d2}$ determines the EMI injection on $v_{ref}$ at high frequencies. Taking into account the closed loop behavior, it can be concluded that the EMI ripple on $v_{ref}$ is maximal at EMI frequencies lying close to the GBW of the OTA [2].

### 2.1.2 Large signal analysis

Previously, the transistors were simplified by linear models. This is representative as long as the EMI amplitude is small. However, as the injected EMI level increases, small signal analyses are not valid any longer, since they do not include nonlinear phenomena like the ubiquitous DC shift. In order to predict the

[2] EMI frequencies lying below the GBW of the OTA are significantly attenuated by the loop gain, while those situated high above the GBW of the OTA are filtered by decoupling capacitor $C_{d2}$ (as well as by optional capacitors decoupling the power supply).

latter, the drain current of $M_1$ is averaged over time. Refer to Fig. 6.3a. The drain-source current of $M_1$ ($I_{d1}$) is equal to the source-drain current of $M_3$ ($I_{d3}$) when there is no EMI, namely:

$$I_{d1\_noEMI} = I_{d3\_noEMI} = \frac{\mu C_{ox}}{2} \cdot \frac{W}{L} \cdot (V_{SG3} - |V_{tp}|)^2 \tag{6.6}$$

Assume now that EMI is superposed on the power supply, as depicted in Fig. 6.3a. In the event that the transistors in the OTA are not forced out of their saturation region by the EMI injection, and as long as first order models are valid, the average drain current of $M_1$ over time is represented as follows:

$$\begin{aligned}\overline{I_{d1\_withEMI}} &= \lim_{T\to\infty} \frac{1}{T} \int_{\frac{-T}{2}}^{\frac{T}{2}} I_{d1}(t) \cdot d(t) \\ &= \overline{\frac{\mu C_{ox}}{2} \cdot \frac{W}{L} \cdot (V_{SG3} + v_{sg3} - |V_{tp}|)^2} \\ &= I_{d1\_noEMI} + \overline{\frac{\mu C_{ox}}{2} \cdot \frac{W}{L} \cdot (v_{sg3})^2}\end{aligned} \tag{6.7}$$

Where $v_{sg3}$ represents the AC source-gate voltage of $M_3$. Clearly, EMI increases the average value of $I_{d1}$. This in turn, forces the gate voltage of $M_5$ to a higher voltage, which boosts the average value of $V_{ref}$.

For larger EMI levels, the output resistances of $M_1$ and $M_2$ do no longer behave linearly, because of the considerable voltage swings which are taking place at their respective drains. Refer to Fig. 6.5. The positive EMI swings generate velocity saturation in the input pair transistors: velocity saturation tends to decrease the large signal resistance of $M_1$ and $M_2$ [3]. Assuming that the drain of $M_2$ lies at a higher voltage than the drain of $M_1$, the decrease of the large signal output resistance owing to velocity saturation is more pronounced for the former than for the latter. Conversely, the negative EMI swings force the input pair transistors $M_1$-$M_2$ into their linear region, which increases the large signal output resistance of $M_1$-$M_2$. This time, however, the increase of the large signal output resistance is predominant for $M_1$, since in this particular case, this transistor lies closer to the linear region owing to its lower drain voltage. Consequently, expression (6.7) does not represent the correct average drain current of $M_1$ any longer: as a result, when the EMI level increases, the average drain current $\overline{I_{d1}}$ steadily decreases while $\overline{I_{d2}}$ increases by the same amount, since the sum of both average currents remains equal to the tail current $I_{b1}$. These combined effects decrease the average gate voltage of $M_5$ and

[3] The large signal output resistance of transistor $M_i$ is defined as the ratio of the mean drain-source voltage ($\overline{V_{dsi}}$) to the mean drain current ($\overline{I_{di}}$).

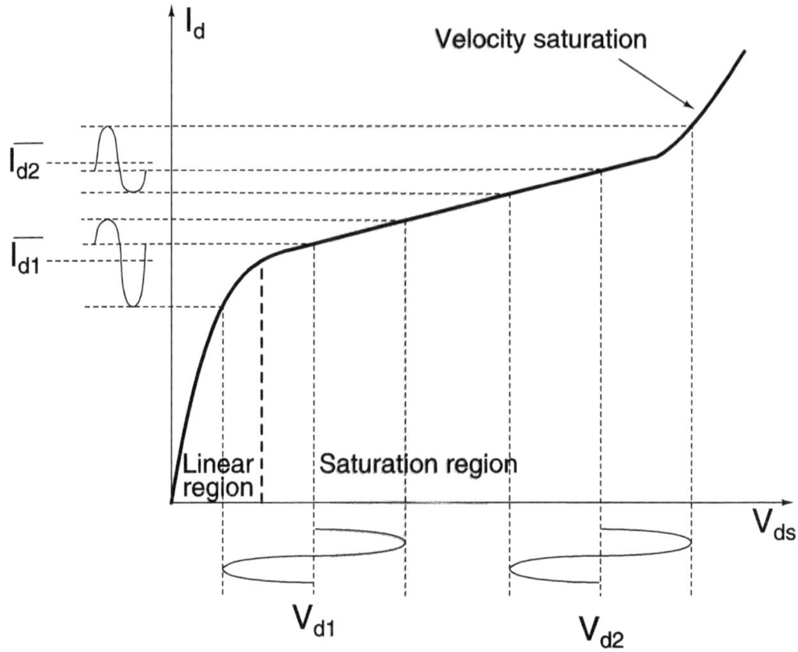


*Figure 6.5.* Representation of the effect of large EMI injection levels on the average drain currents of $M_1$ and $M_2$: observe that $\overline{I_{d1}}$ decreases while $\overline{I_{d2}}$ increases.

consequently $V_{ref}$ at high EMI levels. This effect is illustrated with a simulation example in the upper plot of Fig. 6.6.

First, it will be examined how the gate-source voltage of pass transistor $M_5$ can be kept at a constant value, in order to cancel the DC shift appearing at smaller EMI levels (as derived in (6.7)). Then, the attention will be turned towards masking the input differential pair from the EMI source, in order to shield its nonlinear output resistance at very high EMI levels, while providing an equal drain voltage to $M_1$ and $M_2$. The former issue is addressed first in the next section.

## 2.2 EMI resisting Kuijk bandgap reference (PPD)

A first improvement is realized by replacing the NMOS pass transistor $M_5$ by a PMOS transistor and by connecting the compensation capacitance $C_{d1}$ between the gate and the source of $M_5$ (Fig. 6.7). This topology is referred to as the PMOS pass device topology (PPD). Compensation capacitor $C_{d1}$ couples the EMI to the gate of $M_5$: this way, $M_5$ has a constant gate-source voltage, meaning that, disregarding parasitic effects and considering a first order transistor model, the drain current sourced by $M_5$ remains constant. Because of this, the coupling of EMI to the reference voltage is strongly reduced at low

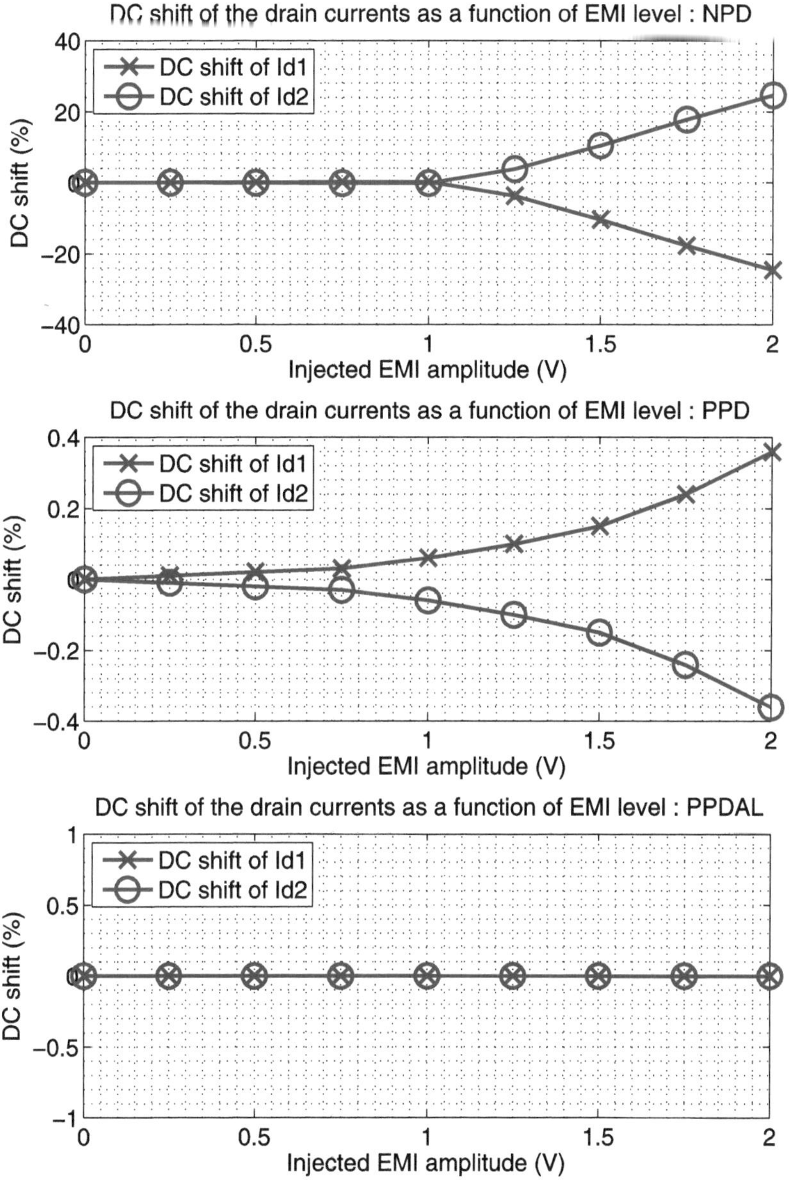


*Figure 6.6.* Simulation of the DC shift of the drain currents of $M_1$ ($I_{d1}$) and $M_2$ ($I_{d2}$) as a function of the injected EMI level for the NPD (upper plot), PPD (middle plot) and PPDAL (lower plot) bandgap topologies. The EMI frequency is equal to 100 MHz.

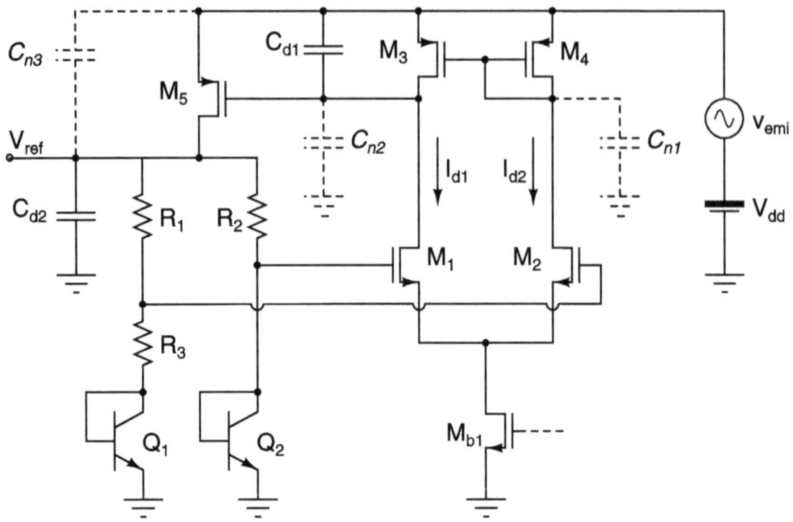


*Figure 6.7.* EMI resisting Kuijk bandgap voltage reference circuit: basic schematic with a PMOS pass device and compensation capacitor between its gate and source (PPD).

EMI frequencies. Again, a distinction is made between small and large EMI amplitudes: the small signal analysis is discussed first [4].

### 2.2.1 Small signal analysis

Applying the same basic abstractions as formulated in Sect. 2.1.1, the open loop transfer function between the EMI source and the reference voltage output node is calculated. As previously, this yields a second order expression, expressed as follows:

$$A_{DD2}(s) = \frac{v_{ref}(s)}{v_{emi}(s)} = \frac{2 \cdot R_L}{r_o} \cdot \frac{(\frac{s}{\omega_{z1}}+1)\cdot(\frac{s}{\omega_{z2}}+1)}{(\frac{s}{\omega_{p1}}+1)\cdot(\frac{s}{\omega_{p2}}+1)} \tag{6.8}$$

The poles and zeros are now given by:

$$\begin{cases} \omega_{p1} = \frac{2}{C_{d1}\cdot r_o}, & \omega_{p2} = \frac{1}{C_{d2}\cdot R_L} \\ \omega_{z1} = \frac{1}{C_{n3}\cdot r_o}, & \omega_{z2} = \frac{4}{r_o\cdot C_{d1}} \end{cases} \tag{6.9}$$

Similarly to what has been observed when studying the NPD structure, the dominant pole of $A_{DD2}(s)$ ($\omega_{p1}$) coincides with the dominant pole of the

[4] Observe that the stability of the PPD structure is not guaranteed: depending on the current flowing thorough the bandgap cell, the impedance at the output node may be very high, which imposes a limitation on capacitor $C_{d2}$ decoupling this node. In present designs, the bandgap cell has a small signal impedance of 2.9 kΩ, while $C_{d1}$ and $C_{d2}$ are both equal to 8 pF (refer to Table 6.1 further on), meaning that the stability requirements are easily fulfilled.

opamp, and the ratio between $C_{n3}$ and $C_{d2}$ determines the EMI injection at higher frequencies. However, the magnitude of $A_{DD2}(s)$ at DC and at lower EMI frequencies is much lower compared to $A_{DD1}(s)$, because of the constant gate-source voltage of pass transistor $M_5$. This effect is clearly observed in the Bode plot depicted in Fig. 6.4. This implies that the EMI injection to $V_{ref}$ is highly attenuated: the EMI induced ripple is hereby strongly reduced.

### 2.2.2 Large signal analysis

Ideally, the PPD bandgap circuit is free of DC shift, since the gate and the source of $M_5$ are shorted to each other by means of capacitor $C_{d1}$. However, at higher EMI frequencies, the gate of $M_5$ is not perfectly equal to its source, because of the capacitive division taking place between $C_{d1}$ and the parasitic capacitance $C_{n2}$. In the same sense, capacitor $C_{d1}$ forms a high impedance at lower EMI frequencies, and consequently, does not couple EMI readily to the gate of pass transistor $M_5$. Finally, for high EMI levels, the EMI which is coupled to the gate of $M_5$, appears in full across the drains of $M_1$ and $M_2$. As described previously, the unequal nonlinear output resistances of $M_1$ and $M_2$ generate a different DC shift in the average drain currents of both transistors. This effect is illustrated with a simulation example in the middle plot of Fig. 6.6, using the same OTA as in the previous section: observe that the improvement compared to the NPD structure is roughly an order of magnitude. Therefore, a circuit solution shielding the drain of $M_1$ and $M_2$ while forcing the DC currents in both opamp branches to equal values must be designed. Additionally, the fact that the gate of $M_5$ may be biased at voltages lying below the generated reference voltage, reduces DC shift on $V_{ref}$ owing to clipping of $M_5$, which may occur during the negative swings of very large EMI levels.

## 2.3 EMI resisting Kuijk bandgap reference (PPDAL)

Since the nonlinear output resistance of the input pair is the main cause of DC shift in the PPD topology, a NMOS current mirror is added at the drain of input pair $M_1$-$M_2$. This topology is identified as the PMOS pass device with active load (PDDAL) topology and is depicted in Fig. 6.8. Current mirror $M_6$-$M_7$ masks the nonlinear output resistances of $M_1$-$M_2$ by using positive feedback [Red08b]. This property is described here below. Consider the structure depicted in Fig. 6.9a [Red08b]. Output impedance $Z_{out}$ seen in the output node $V_{out}$ is calculated using small signal analysis:

$$Z_{out} = \frac{r_{o3}}{r_{o3} \cdot \left(\frac{R_2 - R_1}{R_1 \cdot R_2}\right) + 1} \tag{6.10}$$

Where $r_{o3}$ is the output resistance of $M_3$. Based on (6.10), this structure can therefore be used to generate negative impedances, by changing the ratio of $R_1$ and $R_2$. More importantly, in the particular case that $R_1$ and $R_2$ are equal

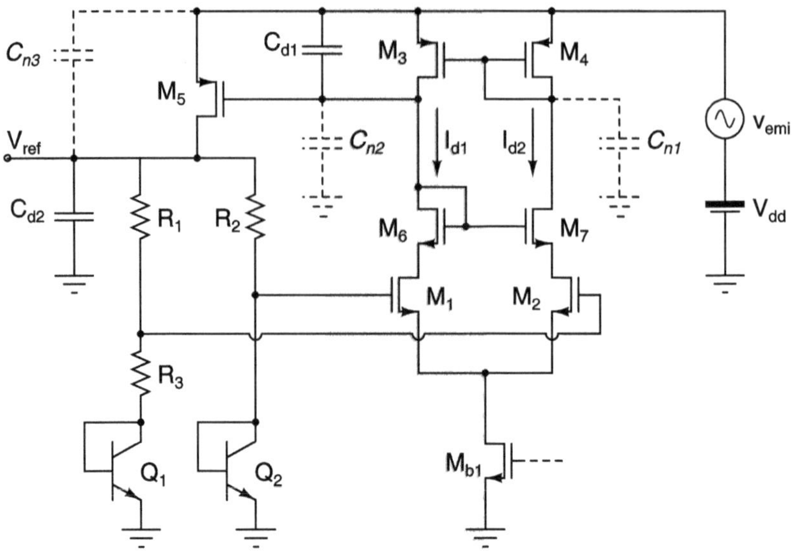


*Figure 6.8.* EMI resisting Kuijk bandgap voltage reference circuit: basic schematic with a PMOS pass device, rerouted compensation capacitor and EMI shielding active load (PPDAL).

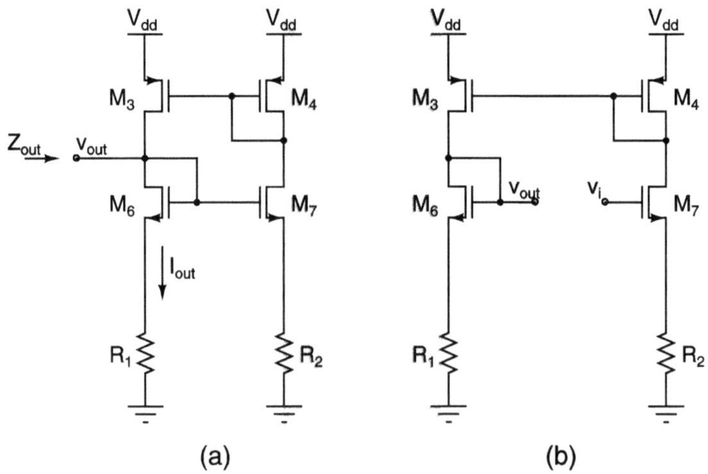


*Figure 6.9.* Active load using positive feedback (a) Principle. (b) Open loop.

to each other, the output impedance of this small structure is equal to $r_{o3}$, and this independently of the absolute values of $R_1$ and $R_2$. In other words, $R_1$ and $R_2$ are masked by the positive feedback loop formed by $M_6$ and $M_7$. In the event that this structure is used as an active load for a differential pair, $R_1$ and $R_2$ represent the output resistances of the input pair transistors. The operation of this circuit is explained as follows: refer to the small signal equivalent open loop circuit depicted in Fig. 6.9b. The circuit contains a positive feedback loop, since if the gate voltage of $M_7$ increases, the gate voltage of $M_3$ and $M_4$ decreases, and consequently, the gate voltage of $M_6$ increases. Positive feedback loops are potentially unstable, and it must be verified that the open loop gain of this positive feedback loop stays below unity. In the event that $g_{m6}$ is equal to $g_{m7}$ and that $R_1$ and $R_2$ are identical, the following expression for the open loop gain is derived:

$$\frac{v_{out}}{v_i} = \frac{r_{o7}}{R_2 + r_{o7}} \tag{6.11}$$

Clearly, as long as $R_2$ stays above zero, the loop gain remains below unity, meaning that the circuit is always stable. Up till now, $R_1$ and $R_2$ were considered to be identical: however, in practice, there is always some mismatch between $R_1$ and $R_2$, impacting the behavior of this structure. In addition, as illustrated in Fig. 6.5, the equivalent output resistances of $M_1$ and $M_2$ differ significantly from each other when large EMI levels are injected in their respective drains, owing to their unequal drain voltages. The question is how this reflects on the impedance masking property of this circuit. Consider that $R_2 = R_1 + \Delta R_1$, where $\Delta R_1$ represents the absolute mismatch between $R_1$ and $R_2$. Equation (6.10) is then rewritten as follows:

$$Z_{out} = \frac{r_{o3}}{r_{o3} \cdot \left(\frac{\Delta R_1}{R_1 \cdot (R_1 + \Delta R_1)}\right) + 1} \tag{6.12}$$

Provided that the following inequality holds:

$$\frac{|\Delta R_1|}{R_1} \ll \frac{R_1}{r_{o3}} \tag{6.13}$$

Then (6.12) is simplified to:

$$Z_{out} \approx r_{o3} \tag{6.14}$$

Which means that as long as the mismatch term is smaller than the ratio between $R_1$ and $r_{o3}$, the output impedance of this structure is equal to $r_{o3}$, regardless of the absolute value of $R_1$ and $R_2$. When used as an active load for a differential pair, $R_1$ and $R_2$ are high impedances lying in the range of $r_{o3}$, and so the condition expressed in (6.13) is quite easily fulfilled under normal circumstances.

This ingenious little circuit has a major impact on the EMI immunity of PPDAL Kuijk bandgap voltage reference circuits. As usual, the small and large signal behavior is discussed separately.

### 2.3.1 Small signal analysis

Applying the same basic abstractions as formulated in Sect. 2.1.1, the transfer function from the EMI source to the reference voltage output node is expressed as follows (refer to Fig. 6.8):

$$A_{DD3}(s) = \frac{v_{ref}(s)}{v_{emi}(s)} = \frac{R_L}{r_o} \cdot \frac{(\frac{s}{\omega_{z1}}+1)\cdot(\frac{s}{\omega_{z2}}+1)}{(\frac{s}{\omega_{p1}}+1)\cdot(\frac{s}{\omega_{p2}}+1)} \tag{6.15}$$

The poles and zeros are equal to:

$$\begin{cases} \omega_{p1} = \frac{1}{C_{d1}\cdot r_o}, & \omega_{p2} = \frac{1}{C_{d2}\cdot R_L} \\ \omega_{z1} = \frac{1}{C_{n3}\cdot r_o}, & \omega_{z2} = \frac{2}{r_o\cdot C_{d1}} \end{cases} \tag{6.16}$$

Observe that, the dominant pole frequency of $A_{DD3}$ ($\omega_{p1}$) as well as the opamp pole frequency have been halved. Conversely, the opamp voltage DC gain has doubled, which decreases the closed loop transfer function even more at low EMI frequencies. This is clearly illustrated in the Bode plot depicted in Fig. 6.4.

### 2.3.2 Large signal analysis

In addition to the advantages which were achieved by using a PMOS pass transistor and by shorting its AC gate-source voltage by means of compensation capacitor $C_{d1}$, two major benefits can be expressed in view of large EMI signals, when the PPDAL and PPD bandgap structures are compared to each other:

- First of all, the output resistances of differential pair transistors $M_1$-$M_2$ are masked from the EMI source: indeed, the OTA output impedance is equal to $r_{o3}$ owing to the positive feedback, and this independently of the output resistances of $M_1$-$M_2$, as derived in (6.14).
- Secondly, the drain voltages of input pair transistors $M_1$-$M_2$ are kept at an equal value. This means that less DC shift is taking place because the systematic offset arising from unequal drain voltages of $M_1$-$M_2$ is cancelled.
- Thirdly, the current mirror formed by $M_6$ and $M_7$ forces the drain currents $I_{d1}$ and $I_{d2}$ to an equal value so that they do not contribute directly to DC shift.

This effect is again illustrated with a simulation example in the lower plot of Fig. 6.6: observe that the DC shift of the drain currents is now close to zero owing to the masking property of $M_6$ and $M_7$. This constitutes a huge improvement compared to the NPD and the PPD structures. There is, however, a possible disadvantage of using this structure, stemming from the fact that the voltage headroom in the OTA is reduced owing to the presence of cascoded transistors $M_6$ and $M_7$. This potential threat must be kept in mind, and verified with simulations.

## 2.4 Startup circuit and biasing

The biasing of current source transistor $M_{b1}$ must be performed carefully, since the tail current may not be influenced by the EMI which is injected in the power supply terminal: for this reason, an EMI resisting current mirror topology has been used (described in Chap. 3). This has been illustrated for the NPD bandgap topology in Fig. 6.10. Observing a careful dimensioning of $M_{b1}$, $M_{b2}$, $M_{b3}$ and $M_{b4}$, and of $C_{b1}$ and $C_{b2}$, a high EMI suppression between $V_{dd}$ and the gate of $M_{b1}$ is achieved this way. In addition, Kuijk bandgap circuits have a parasitic operating point at zero volt, and consequently, they are not self-starting. For this reason, a startup circuit was foreseen internally, by means of transistor $M_{su}$. The operation of this transistor is explained referring to Fig. 6.10: in case the voltage at $V_{ref}$ is small or close to zero, the gate of $M_3$ is pulled down by $M_{su}$, which in turn increases the voltage at the gate of NMOS pass transistor $M_5$. Consequently, this forces $V_{ref}$ to a higher value. Once $V_{ref}$ equals the wanted bandgap voltage, transistor $M_{su}$ turns off, and has no further use in the circuit.

## 2.5 Measurements

The three discussed Kuijk bandgap reference circuits (NPD, PPD and PPDAL) were integrated on a test-IC which was processed in the AMIS 0.35 μm tech-

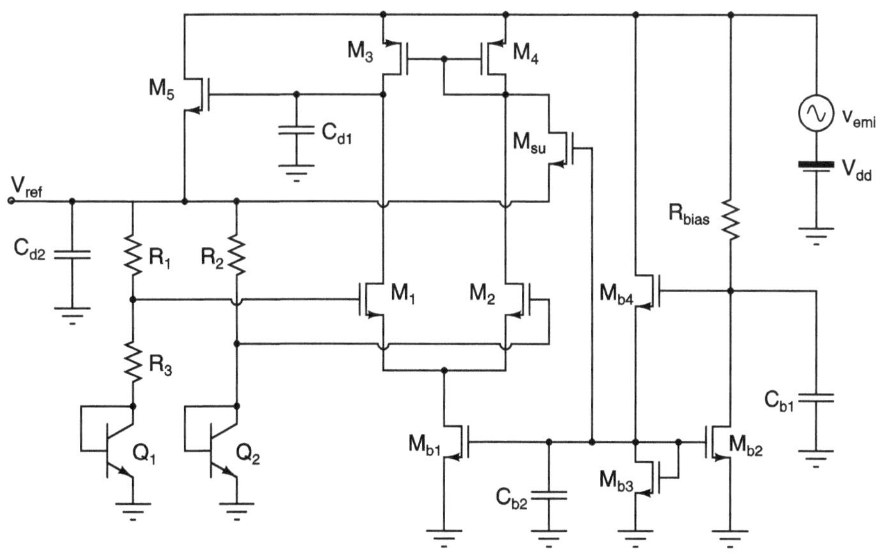


*Figure 6.10.* Kuijk NPD bandgap voltage reference circuit with biasing of tail current transistor $M_{b1}$ and startup circuit.

*Table 6.1.* Design values.

| | *NPD* | *PPD* | *PPDAL* |
|---|---|---|---|
| $M_1$-$M_2$ | 90 μm/2.5 μm | 90 μm/2.5 μm | 25 μm/0.5 μm |
| $M_3$-$M_4$ | 60 μm/3 μm | 60 μm/3 μm | 60 μm/3 μm |
| $M_6$-$M_7$ | – | – | 25 μm/0.5 μm |
| $I_{b1}$ | 21.5 μA | 21.5 μA | 21.5 μA |
| $M_5$ | 30 μm/0.5 μm | 60 μm/0.5 μm | 60 μm/0.5 μm |
| $r_{o5}$ | 43 kΩ | 67 kΩ | 67 kΩ |
| $C_{b1}, C_{b2}, C_{d1}, C_{d2}$ | 8 pF | 8 pF | 8 pF |
| $R_L$ | 2.9 kΩ | 2.9 kΩ | 2.9 kΩ |
| $A_{DC\_OTA}$ | 50 dB | 50 dB | 61 dB |
| OTA GBW | 3.6 MHz | 3.6 MHz | 3.6 MHz |
| OTA phase margin | 81° | 81° | 82° |

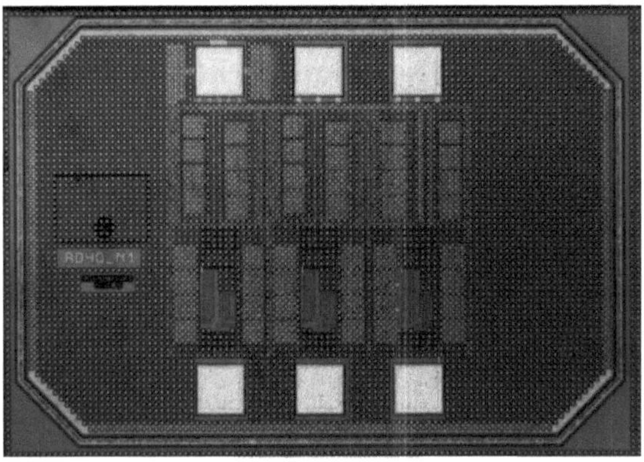

*Figure 6.11.* Microphotograph of the bandgap test-IC. The size of the circuit is approximately 0.6 mm × 0.5 mm.

nology [5]. A microphotograph of this test-IC is shown in Fig. 6.11. The transistor design values are summarized in Table 6.1. The EMI source was injected in the power supply through a resistor of 100 Ω according to the DPI standard measurement setup [DPI]: the EMI frequency ranges between 150 kHz and 1 GHz. The measured DC shifts of the respective reference voltages of the NPD, PPD and PPDAL bandgap structures are plotted in Fig. 6.12 for an EMI power level of 4 dBm, which corresponds to an EMI source amplitude of 1 V.

[5] Processed in AMIS I3T50, using the low-voltage components [Pes03].

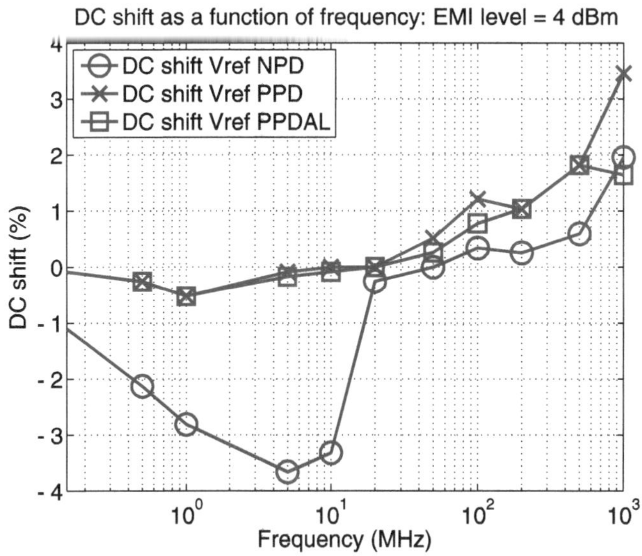


*Figure 6.12.* DC shift as a function of the EMI frequency, for the NPD, PPD and PPDAL bandgap circuits: EMI level = 4 dBm (1 V amplitude).

Observe in this plot that the DC shift generated in the classic NPD circuit is very high at low EMI frequencies, compared to the PPD and the PPDAL structures. At higher EMI frequencies, the DC shift of the reference voltage associated to the PPD structure worsens, starting from approximately 100 MHz: this is explained by the EMI injection which is taking place at the drain of $M_1$-$M_2$. The same trend is observed for the classic NPD structure, albeit being less pronounced since the EMI injection is stronger in the PPD case (it is directly coupled through capacitor $C_{d1}$, as depicted in Fig. 6.7). Conversely, the DC shift in the PPDAL bandgap circuit stays relatively small over the full EMI frequency range because $M_1$-$M_2$ are shielded by $M_6$-$M_7$ (Fig. 6.8). Adding a small internal decoupling capacitor in the power supply would improve this situation considerably at high EMI frequencies: however, this decoupling capacitor would be ineffective at low EMI frequencies, which are therefore more critical.

Next, the injected EMI power level has been increased up to 10 dBm, in order to comply with the DPI measurement for a zone 3 appliance (refer to Chap. 2 and [DPI]): this corresponds to an EMI voltage amplitude of 2 V. The resulting measurement has been depicted in Fig 6.13. Clearly, the reference voltage generated by the NPD bandgap circuit is unrecognizable for such high EMI injected levels: this is caused by the fact that such high EMI levels on the power supply periodically debias the circuit, and consequently, the correct DC level at the gate of the NMOS pass transistor can not be maintained any longer.

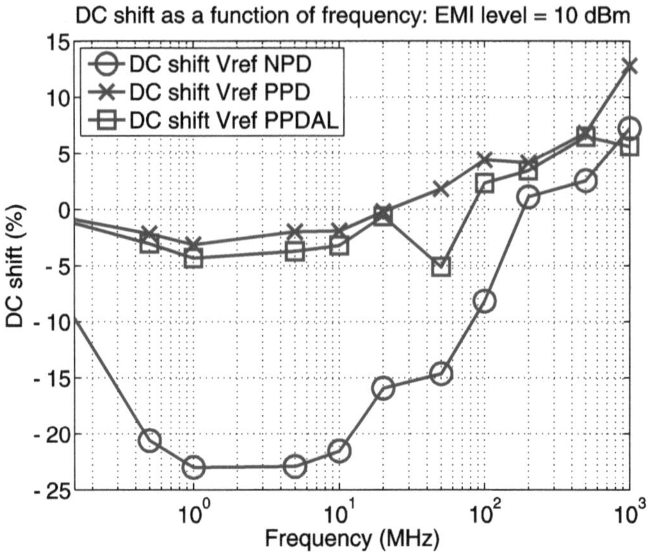


*Figure 6.13.* DC shift as a function of the EMI frequency, for the NPD, PPD and PPDAL bandgap circuits: EMI level = 10 dBm (2 V amplitude).

Moreover, the reference voltage generated by the PPDAL circuit is slightly worse than the one produced by its PPD counterpart at low EMI frequencies. This is due to the fact that the PPD circuit uses less cascoded devices, and so it can therefore remain much longer in its operating region. However, the DC shift of the PPD circuit quickly increases for higher EMI frequencies, whereas the PPDAL proves again to be less susceptible at higher EMI frequencies.

Figures 6.14 and 6.15 depict the measured EMI induced peak-to-peak ripple on the generated reference voltages of the designed NPD, PPD and PPDAL bandgap structures, as a function of the EMI frequency for the considered EMI injection levels of respectively 4 dBm and 10 dBm. Clearly, the ripple corresponding to the PPD and PPDAL structures is considerably smaller than the one associated to the NPD circuit. However, in view of the previously performed small signal analyses and their respective Bode magnitude plots depicted in Fig. 6.4, a more significant improvement may at first sight be expected: this fact is clarified as follows. The drain area of PMOS pass transistor $M_5$ in the PPD and the PPDAL bandgap structure has been minimized, in order to reduce its parasitic drain-bulk capacitance (represented by a generic capacitor $C_{n3}$ in Figs. 6.7 and 6.8): however, this reduces its AC output resistance. Considering that the approximate small signal resistance of the bandgap cell itself ($R_L$) is equal to 2.9 k$\Omega$, the voltage division realized between $V_{dd}$ and $V_{ref}$ owing to $R_L$ and $r_{o5}$, yields a peak to peak ripple at low EMI frequencies which is approximately equal to 83 mV (EMI level = 4 dBm) and

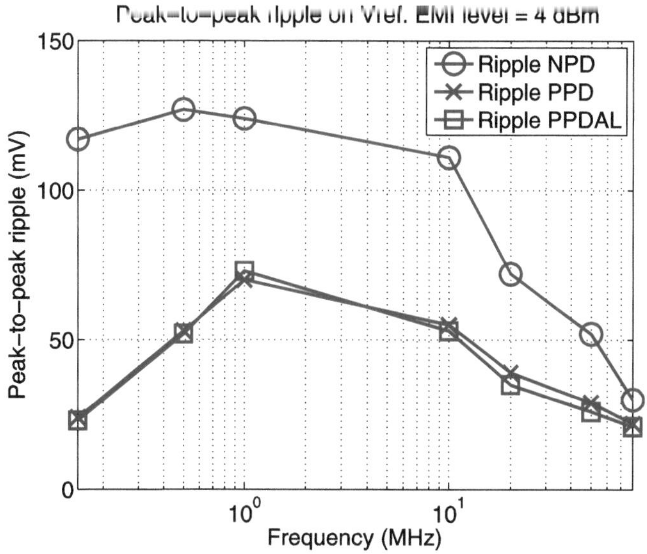


*Figure 6.14.* Peak-to-peak ripple on $V_{ref}$ as a function of the EMI frequency, for the NPD, PPD and PPDAL bandgap circuits: EMI level = 4 dBm.

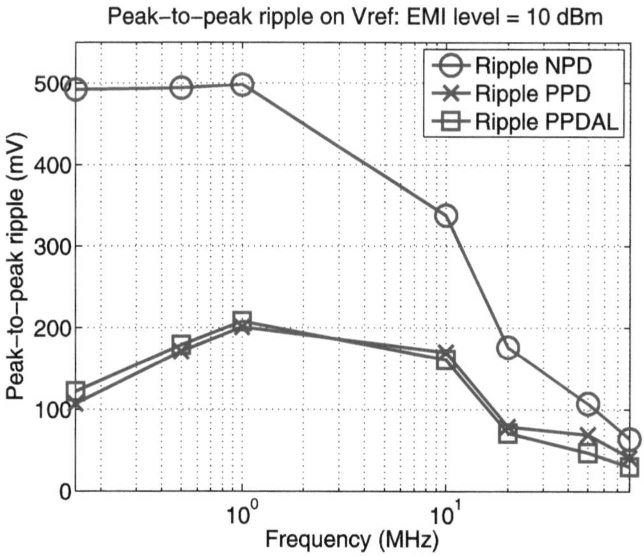


*Figure 6.15.* Peak-to-peak ripple on $V_{ref}$ as a function of the EMI frequency, for the NPD, PPD and PPDAL bandgap circuits: EMI level = 10 dBm.

166 mV (EMI level = 10 dBm). These theoretical values lie very close to the ones measured at low EMI frequencies (for instance at 1 MHz), as can be seen on Figs. 6.14 and 6.15.

# 3 Case study 2: EMI resisting low dropout voltage regulators

Owing to stringent power management constraints in modern integrated systems, low dropout (LDO) voltage regulators are required to have a small voltage drop across the pass device [Mil07]. LDO voltage regulators are prevalently used in order to derive a fixed low-voltage power supply from a constantly varying external supply line, even when the difference between both of them drops below 0.5 V. For instance, in automotive applications, the power supply of onboard electronic devices must be guaranteed even though the battery voltage may drop to about 6 V during starting [Bon01]. Additionally, since LDO regulators supply the necessary power to various subcircuits in large IC's, they must remain as immune as possible to EMI which is injected in the supply rails, as has been highlighted in the introductory section of this chapter. This case study examines a possible design of a LDO voltage regulator, having an intrinsic improved immunity against EMI which is conveyed on the power supply terminal. This circuit is inspired by the observations which were developed throughout the first case study. First, using classic LDO regulator structures as a starting point, the EMI resisting LDO regulator structure is introduced and described in detail. A theoretical analysis illustrates that the major EMI coupling in LDO regulator topologies is caused by the pass device, as well as by the finite PSRR of the OTA. A design example compares the EMI rejection of a classic LDO regulator versus the new EMI resisting structure, hereby demonstrating the latter's superior performance.

## 3.1 EMI issues in LDO voltage regulator circuits

Generally speaking, LDO regulators driving an arbitrary resistive load (represented by resistor $R_L$) can be categorized according to the pass device, which is either a NMOS (common-drain LDO regulator, Fig. 6.16a) either a PMOS transistor (common-source LDO regulator, Fig. 6.16b) [Kay07, San06]. Factor $\alpha$ expresses the ratio between the wanted output voltage ($V_{out}$) and the given bandgap reference voltage ($V_{ref}$) [6]. Because of the ever increasing power efficiency constraints, preference is given to common-source LDO regulators since these exhibit a much smaller voltage drop compared to their common-drain counterparts. However, a considerable disadvantage of common-source LDO regulators is the presence of a second low frequency pole, associated to the output node of the voltage regulator ($V_{out}$), which may seriously jeopardize

[6] This factor $\alpha$ is very often implemented using a high precision resistive divider.

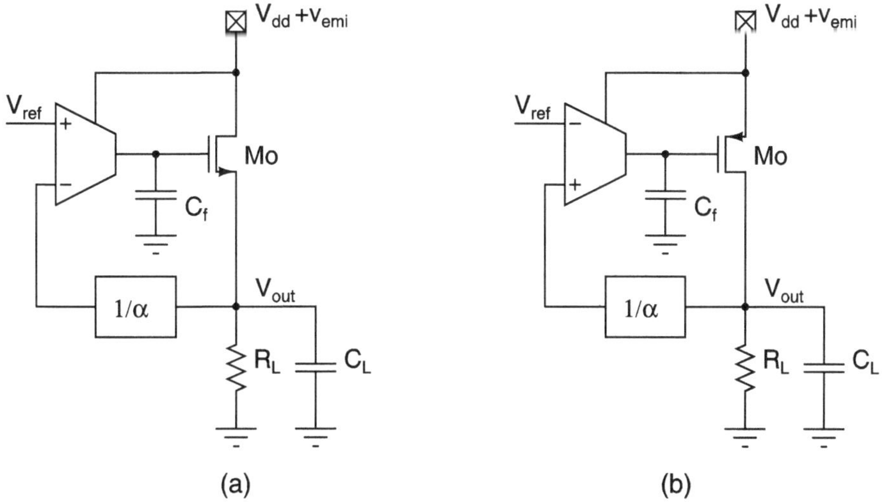


*Figure 6.16.* (a) Classic common-drain LDO voltage regulator topology. (b) Classic common-source LDO voltage regulator topology.

the overall stability of the regulator circuit [Bes98, San06]. More importantly, since the source of PMOS pass transistor $Mo$ is connected to the positive power supply, any EMI disturbance which is present on $V_{dd}$ propagates to the output node $V_{out}$.

In the event that a common-drain LDO regulator topology is used, the EMI immunity of the pass device increases, while a smaller voltage drop across the latter can be achieved by boosting the gate voltage [Giu06]. However, the charge pump circuit which is used for this purpose is commonly very susceptible to EMI which is superposed on the power supply voltage: this increases the EMI coupling to the gate of $Mo$, and consequently, to $V_{out}$. In the same way, owing to the finite PSRR of the OTA, the disturbances present on the supply line couple to the gate of $Mo$. Consequently, both LDO regulator types present a non-negligible susceptibility to EMI which is injected in the power supply rail. Consider the topology which is depicted in Fig. 6.17a. A PMOS pass transistor is used, but this time, the compensation capacitance of the OTA ($C_f$) is connected between $V_{dd}$ and the gate of $Mo$. This way, the gate-source voltage of $Mo$ is kept constant, meaning that the drain current through $Mo$ remains theoretically unaffected by the injected EMI level. This approach is similar to the one used in Sect. 2.2 in order to improve the electromagnetic immunity of classic Kuijk bandgap reference circuits. Furthermore, the finite PSRR of the OTA is no longer an issue, since EMI is directly coupled to the gate of $Mo$ through $C_f$. It will now be examined mathematically how this circuit must be dimensioned to ensure a stable operation, and how a high EMI immunity is

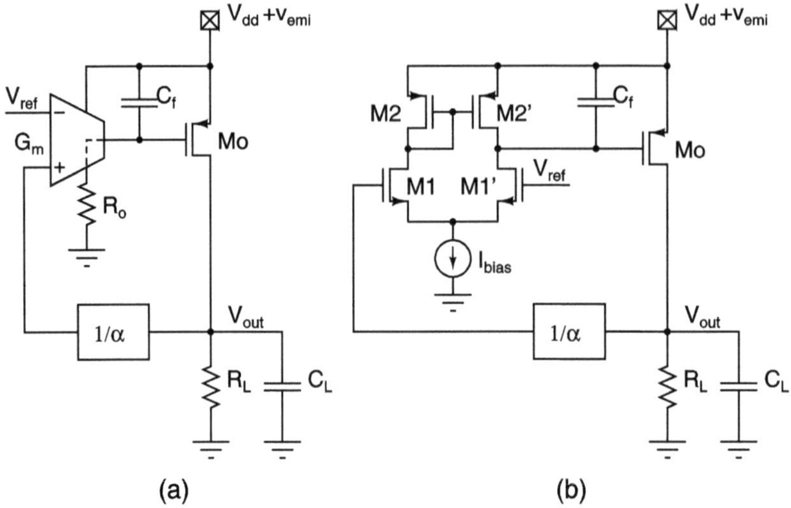


*Figure 6.17.* (a) EMI resisting LDO voltage regulator topology. (b) EMI resisting LDO voltage regulator topology, with detailed OTA schematic.

achieved. It should be observed at this point that two possible strategies are possible in view of approaching the stability considerations in this circuit.

The first possibility is to consider the output pole (present at node $V_{out}$) as being the dominant one: this approach is relevant when $C_L$ is dimensioned much larger than $C_f$. This is the case when $V_{out}$ is an external node, which can be coupled to an large external decoupling capacitor. Such a situation is fortunate in view of improving the EMC behavior, since this means that the EMI reaching node $V_{out}$ is extremely attenuated by this decoupling capacitor: additionally, since $V_{gs1}$ is kept constant, the DC shift is virtually eliminated at low EMI frequencies, as seen previously. Consequently, EMC problems are less likely to be expected in this situation. In the second event, the output pole created at node $V_{out}$ is not dominant: this may for instance be the case when $V_{out}$ is an internal node, which is decoupled internally. Owing to this smaller decoupling, EMC problems are more likely to occur, and this situation is examined next. First of all, the DC open loop gain of the LDO regulator is expressed as follows:

$$A_{DC} = \frac{G_m \cdot R_o \cdot g_{mo} \cdot R_L}{\alpha} \tag{6.17}$$

where $G_m$, $R_o$ and $g_{mo}$ represent respectively the OTA transconductance and output resistance, and the transconductance of $M_o$. Since there are two high impedance nodes, the loop gain transfer function contains two poles:

$$\omega_{p1} = \frac{1}{C_f \cdot R_o}, \qquad \omega_{p2} = \frac{1}{C_L \cdot R_L} \tag{6.18}$$

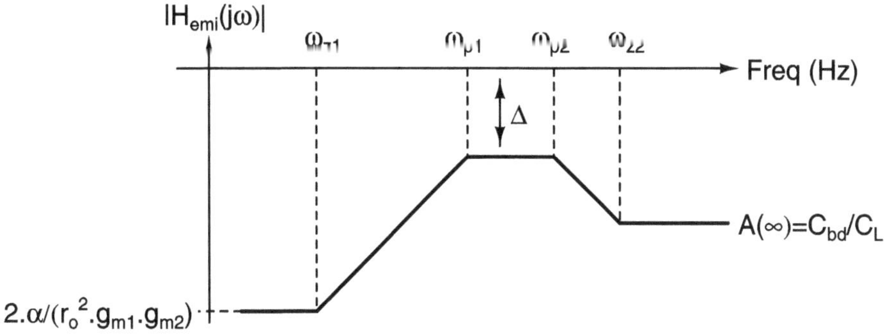


*Figure 6.18.* Bode plot representing the EMI coupling to the output of the EMI resisting LDO structure ($V_{out}$).

The non-dominant pole ($\omega_{p2}$) must lie far enough from the dominant one ($\omega_{p1}$) in order to guarantee stability. This constraint is expressed with factor $\beta$:

$$\omega_{p2} = \beta \cdot GBW \rightarrow \omega_{p2} = \beta \cdot \omega_{p1} \cdot A_{DC} \tag{6.19}$$

As an example, if $\beta$ is equal to 1, then the second pole lies at the gain bandwidth product (GBW) frequency, and the corresponding phase margin is equal to 45° [San06]. Substituting (6.17) and (6.18) in (6.19), results in the following expression for $C_f$:

$$C_f = \frac{\beta}{\alpha} \cdot R_L^2 \cdot C_L \cdot G_m \cdot g_{mo} \tag{6.20}$$

Observing previous relationship, the stability requirement of the proposed LDO voltage regulator is well defined. Once again, the EMI behavior is now examined using a practical one stage OTA, designed as a classic differential pair which is actively loaded by a current mirror (refer to Fig. 6.17b). Calculating the EMI transfer function from $V_{dd}$ to $V_{out}$, assuming that all the output resistances ($r_o$) of all the individual transistors are equal to each other and using the stability requirement described in (6.20), yields the following simplified transfer function:

$$\begin{aligned} H_{emi}(s) &= \frac{V_{out}(s)}{V_{in}(s)} \\ &= \frac{\beta \cdot r_o^2 \cdot R_L^2 \cdot C_{bdo} \cdot C_L \cdot g_{m1} \cdot g_{m2} \cdot s^2 + \beta \cdot r_o \cdot R_L^2 \cdot C_L \cdot g_{m1} \cdot g_{m2} \cdot s + 2 \cdot \alpha}{r_o^2 \cdot g_{m1} \cdot g_{m2} \cdot (\beta \cdot C_L^2 \cdot R_L^2 \cdot s^2 + \beta \cdot C_L \cdot R_L \cdot s + 1)} \end{aligned} \tag{6.21}$$

The poles and zeros are approximated by:

$$\begin{cases} \omega_{p1,p2} = \frac{\beta \pm \sqrt{\beta^2 - 4 \cdot \beta}}{2 \cdot \beta \cdot R_L \cdot C_L} \\ \omega_{z1} = 0, \quad \omega_{z2} = \frac{1}{C_{bdo} \cdot r_o} \end{cases} \tag{6.22}$$

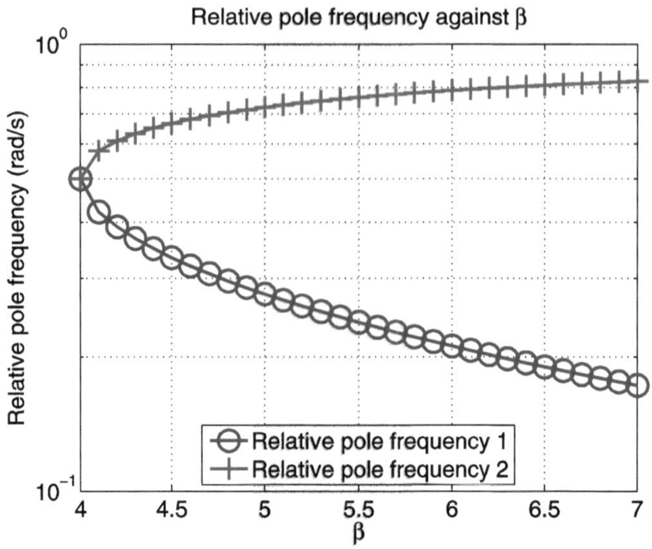


*Figure 6.19.* Relative pole frequencies as a function of $\beta$.

Previous transfer function is represented in the simplified Bode plot depicted in Fig. 6.18. This plot clearly shows that in order to minimize the EMI coupling to $V_{out}$, $\Delta$ must be maximized as much as possible. Firstly, this can be realized by increasing $\beta$, since this reduces the low frequency pole and increases the high frequency pole as can be observed in (6.22). This dependency is depicted in Fig. 6.19. Observe that the highest pole frequency does not increase very much, and consequently, the improvement in reducing $\Delta$ is marginal. Secondly, $\Delta$ is maximized by increasing the output resistance $r_o$, since this reduces the frequency of the highest zero, while the ratio of the parasitic bulk-drain capacitance of $Mo$ ($C_{bd}$) to $C_L$ remains identical. This is clearly seen in the Bode plot in Fig. 6.18, and has been illustrated in Fig. 6.20 using arbitrary design values.

When EMI levels are too large in order to consider linearized transistors, small signal models may not be relied upon, and large signal transistor formulas must be used instead. An ever recurring phenomenon is the EMI induced DC shift, which has been introduced in Chap. 3. Ideally, in this circuit, there is no DC shift, since the gate and the source of $Mo$ are shorted to each other by means of capacitor $C_f$ for AC signals. However, at higher EMI frequencies, the gate of $Mo$ is not perfectly equal to its source, because of the capacitive division taking place between $C_f$ and the parasitic capacitance which is present at the drain of $M_1'$, and because of the nonlinear output resistance of $M_1$-$M_1'$. With this in view, the same shielding circuit as the one introduced in Sect. 2.3, may be used. This alternative has, however, not been investigated further in this work.

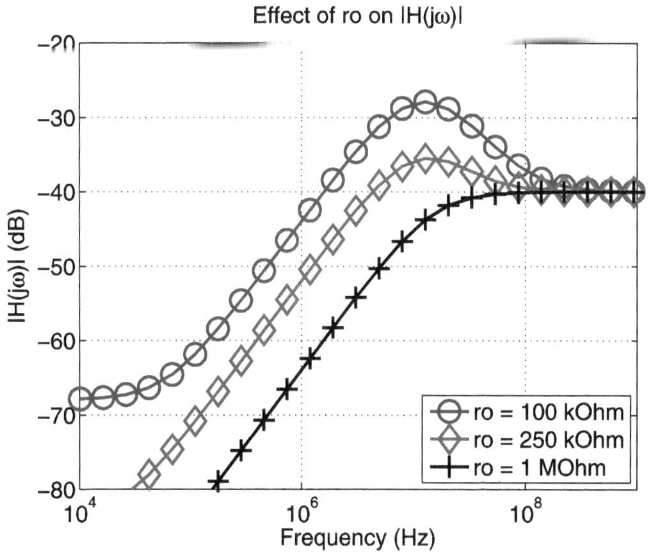


*Figure 6.20.* Effect of output resistance $r_o$ on the magnitude of transfer function $H_{emi}$ ($\alpha = 4$, $\beta = 4$, $C_L = 10$ pF, $C_f = 160$ pF, $R_L = 4$ kΩ, $g_{m1} = g_{m2} = 1$ mS).

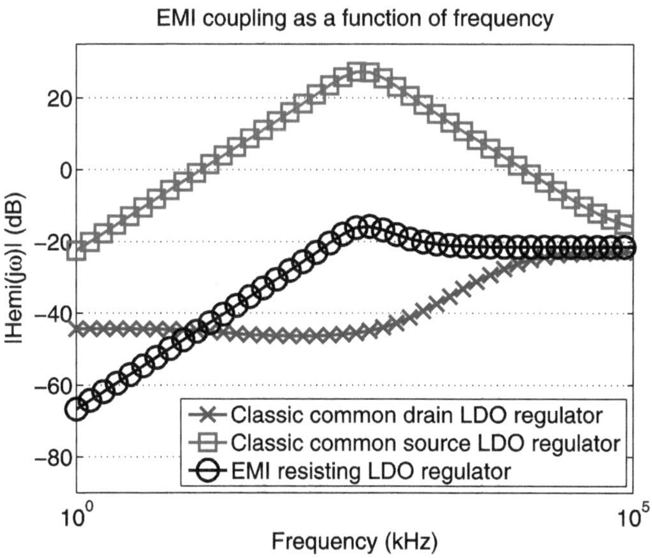


*Figure 6.21.* Circuit simulation results of the EMI coupling to the LDO regulator output node ($V_{out}$).

*Table 6.2.* Design values.

| | *Common-drain LDO regulator* | *Common-source LDO regulator* | *EMI resisting LDO regulator* |
|---|---|---|---|
| $V_{dd}$ | 3.3 V | 3.3 V | 3.3 V |
| $V_{out}$ | 2 V | 3 V | 3 V |
| $R_L$ | 20 kΩ | 20 kΩ | 20 kΩ |
| $Mo$ | 2000 μm/4 μm | 1000 μm/4 μm | 1000 μm/4 μm |
| $M_1$ | 10 μm/4 μm | 10 μm/4 μm | 10 μm/4 μm |
| $M_2$ | 30 μm/2 μm | 30 μm/2 μm | 30 μm/2 μm |
| $I_{bias}$ | 5 μA | 5 μA | 5 μA |
| $C_L$ | 20 pF | 20 pF | 20 pF |
| $C_f$ | 160 pF | 160 pF | 160 pF |

## 3.2 Design example

The three discussed LDO voltage regulator topologies (depicted in Figs. 6.16a, 6.16b and 6.17a) have been simulated using a 0.35 μm CMOS process in order to generate an internal 3 V supply voltage driving a resistive load ($R_L$) of 20 kΩ: the design values are summarized in Table 6.2. The same OTA design (shown in Fig. 6.17b) and the same capacitor values have been used in the three circuits in order to make a fair comparison. Figure 6.21 depicts a simulation of the EMI coupling from $V_{dd}$ to $V_{out}$. Although the common-drain LDO regulator seems to generate the smallest EMI coupling, its NMOS transistor pass device requires a very high drain-source voltage (1.3 V, according to Table 6.2), and can therefore not generate the required 3 V at the output node, making this topology useless in the present case. Observe that the EMI resisting LDO regulator provides the required 3 V, with a maximal coupling of $-16$ dB over the full EMI frequency range. Finally, at high EMI frequencies, transfer function $|H_{emi}(j\omega)|$ converges to the ratio of the parasitic bulk-drain capacitance of $Mo$ ($C_{bdo}$) and $C_L$, as predicted in Fig. 6.18.

Chapter 7

# Epilogue

*Since the 1960s, there have been reports that electromagnetic interference can cause critical-care medical devices to malfunction. Such malfunctions have caused inappropriate therapy, patient injury, mortality, or have had the potential to do so. Fortunately, such incidents are rare, and the incidence of such malfunctions appears to be declining with time. However, vigilance is still required because (1) the electromagnetic compatibility (EMC) of many new radiofrequency (RF) sources and new medical devices being introduced into healthcare is unknown, and (2) there will be a substantial increase in usage of wireless information technology needed by healthcare but the EMC of such technology is unknown.*

—Quoted from [Seg01]

Electronic appliances have fused almost indiscernibly in the pattern of our everyday lives, and the number of EMC incompatibilities associated to the omnipresence of these novel applications is continuously increasing. An indisputable testimony to this fact are the even more stringent EMC related specifications to which actual products must comply before they are released on the consumer market. Adapted legislation regulates the tolerated electromagnetic immunity and emission levels and spectra in electric and electronic appliances, and as a consequence, product manufacturers request IC's from the chip industry, which are capable of resisting the most severe EMC conditions. Owing to the inherent safety considerations, the automotive, aeronautic and medical industries are amongst others particularly keen on developing and using circuits with an increased immunity to EMI, while minimizing as much as possible the detrimental electromagnetic emission.

Clearly, the countering of adverse EMC phenomena taking place at integrated circuit level remains a vast and important research subject, which is paradoxically still somewhat overshadowed by the extensive amount of EMC

J.-M. Redouté, M. Steyaert, *EMC of Analog Integrated Circuits*,
Analog Circuits and Signal Processing,
DOI 10.1007/978-90-481-3230-0_7, © Springer Science+Business Media B.V. 2010

related reference material focused at improving the EMC behavior of PCB's. This creates the false impression that all EMC incompatibilities need to be solved outside the IC. Furthermore, the fundamental relationships existing between electromagnetism and EMC, fuel the overall perception that the latter is a complex and even a quasi-mystic subject, which must in practice be solved according to empirical design guidelines and extensive shielding.

A careful PCB design is of paramount importance so as to fulfill the EMC compliance of a complete application: however, as the interfering frequencies tend to increase, the accompanying EMC problems are moved towards smaller distances, like IC pins and bondwires. In that light, EMC robust integrated circuit design is continuously gaining in significance and global recognition: partly because of economic reasons, prohibiting for instance an all-round shielding and an excessive use of external decoupling devices and chokes, and partly because of the very dense level of integration, forcing many different integrated systems to cohabitate and operate satisfactorily in a very close confinement, and sometimes even on the same chip. Therefore, it is just a matter of time before actual on-chip interconnections will need to be monitored closely as well. Evidently, this will impose even more stringent constraints on EMI robust analog integrated circuit design in the nearby future.

The pervading impression that EMC is an obscure research subject is rooted in the fact that for many years, EMC engineers and specialists solved existing EMC problems by means of their own compliancy rules, which were largely based on their individual experience. Since EMC problems manifested themselves mostly after the design of a full product, these specialists were generally called upon to solve existing problems. This situation has changed inexorably over the last few years, because of very fast product design cycles which are imperatively observed in order to maintain a competitive position in rapidly evolving consumer markets. As a direct consequence, the time and money invested into redesigning a full product for achieving EMC compliancy is simply not effective: nowadays, design engineers are forced to take EMC into account during the design phase of the product itself, meaning that the EMC awareness has evolved from an EMC problem solving to an EMC prevention aspect. Moreover, much more people are involved with EMC throughout the product design cycle, which somewhat contributes to its demystification.

Because EMC rules are closely linked and derived from the most basic laws of electromagnetic theory, and because they are taking place on all the possible coupling paths in a particular circuit or product, this subject is (rightly) considered to be quite complex to understand and to analyze: a basic knowledge and insight of these possible interactions is, therefore, definitely required. Present work made an attempt to contribute to this fascinating subject by studying and improving the electromagnetic susceptibility of analog integrated circuits using a top-down approach. To this end, this book has been structured into three main parts, namely EMI entering and disrupting select analog integrated circuits at output pins, at input pins, and through power supply terminals. In each part, case studies have been designed which – starting from a classic circuit topology exhibiting a particular EMC problem – were made able to withstand higher EMI injection levels while continuing to operate in a satisfactory way.

There seems no better way of completing this work by advocating the same wise statement that has already been mentioned in the first chapter of this book [Deu03a]:

*By designing EMC robust and resisting IC's, let's hope that we can keep up with Moore's Law for a long time.*

The authors hope that this work has at least modestly contributed to this goal.

# References


[Ala97] E. Alarcón, E. Vidal, and A. Poveda, "High-frequency response modeling of continuous-time current mirrors", in: *European Conference on Circuit Theory and Design*, pp. 204–209, Hungary, August 1997.

[Alb81] P. Albert and A.-J. Tudesq, *Histoire de la Radio-Télévision*, Que sais-je, Presses universitaires de France.

[Ale08] S. Alexandersson, "Functional safety and EMC for the automotive industry", in: *IEEE International Symposium on Electromagnetic Compatibility*, pp. 1–6, Detroit, USA, August 2008.

[Alo05] W. Aloisi, G. Giustolisi, and G. Palumbo, "Design and comparison of very low-voltage CMOS output stages", *IEEE Transactions on Circuits and Systems I*, vol. 52, no. 8, pp. 1545–1556, 2005.

[App05] K. Appeltans, "LIN transceivers: a case study of automotive design", in: *IEEE ISSCC Digest of Technical Papers*, 2005, ATAC design forum.

[Arc04] B. R. Archambeault, *PCB Design for Real-World EMI Control*, Kluwer Academic, Amsterdam, 2004.

[Arf95] G. B. Arfken and H. J. Weber, *Mathematical Methods for Physicists*, 4th edn., Academic Press, San Diego, 1995.

[Arm07] Compiled by K. Armstrong and A.E. Hutley, *The First 500 Banana Skins*, Nutwood, UK, 2007.

[Arm08] K. Armstrong, "EMC for the functional safety of automobiles—why EMC testing is insufficient, and what is necessary", in: *IEEE International Symposium on Electromagnetic Compatibility*, pp. 1–6, Detroit, USA, August 2008.

[Bau70] G. Baum and H. Beneking, "Drift velocity saturation in MOS transistors", *IEEE Transactions on Electron Devices*, vol. 17, no. 6, pp. 481–482, 1970.

[Bec08] A. Becker-Gomez, L. T. Viswanathan, and T. R. Viswanathan, "A low-supply-voltage CMOS sub-bandgap reference", *IEEE Transactions on Circuits and Systems II*, vol. 55, no. 7, pp. 609–613, 2008.

[Ben06] S. Ben Dhia, M. Ramdani, and E. Sicard, *Electromagnetic Compatibility of Integrated Circuits*, Springer, Berlin, 2006.

[Bes98] G. W. den Besten and B. Nauta, "Embedded 5 V-to-3.3 V voltage regulator for supplying digital IC's in 3.3 V CMOS technology", *IEEE Journal of Solid-State Circuits*, vol. 33, no. 7, pp. 956–962, 1998.



J.-M. Redouté, M. Steyaert, *EMC of Analog Integrated Circuits*,
Analog Circuits and Signal Processing,
DOI 10.1007/978-90-481-3230-0, © Springer Science+Business Media B.V. 2010

[Bon01] G. Bontempo, T. Signorelli, and F. Pulvirenti, "Low supply voltage, low quiescent current, ULDO linear regulator", in: *IEEE International Conference on Electronics, Circuits and Systems*, vol. 1, pp. 409–412, 2001.

[Boy07] A. Boyer, S. Ben Dhia, and E. Sicard, "Characterisation of electromagnetic susceptibility of integrated circuits using near-field scan", *IEE Electronics Letters*, vol. 43, no. 1, pp. 15–16, 2007.

[Car94] T. Carter, "Switch mode power supplies: an EMI engineer's point of view", in: *Conference Record of IEEE Southcon 94*, pp. 295–300, Orlando, USA, March 1994.

[Car95] J. Carpenter, "Understanding electromagnetism – III. Energy flow", *Engineering Science and Education Journal*, vol. 4, no. 6, 1995.

[Car04] R. M. Carlton, "An overview of standards in electromagnetic compatibility for integrated circuits", *Microelectronics Journal*, vol. 35, no. 6, pp. 487–495, 2004.

[Cas04] H. Casier, "Electronic circuits in an automotive environment", in: *International Solid-State Circuits Conference*, Tutorial, San Fransisco, USA, February 2004.

[Cat95a] J. Catrysse, "Measured distortion of the output-waveform of an integrated OPAMP due to substrate noise", *IEEE Transactions on Electromagnetic Compatibility*, vol. 37, no. 2, pp. 310–312, 1995.

[Cat95b] J. Catrysse, A. Sinnaeve, and G. Vandecasteele, "Measured crosstalk on chips using specially designed components", *IEEE Transactions on Electromagnetic Compatibility*, vol. 37, no. 2, pp. 313–315, 1995.

[Cer05] M. Cervi, D. Pappis, T. B. Marchesan, A. Campos, and R. N. do Prado, "A semiconductor lighting system controlled through a LIN network to automotive application", in: *Conference Record of the Industry Applications Conference*, vol. 3, pp. 1603–1608, October 2005.

[Cha98] E. Chaisson and S. McMillan, *Astronomy, a Beginner's Guide to the Universe*, 2nd edn., Prentice Hall, New York, 1998.

[Cha97] J. F. Chappel and S. G. Zaky, "EMI effects and timing design for increased reliability in digital systems", *IEEE Transactions on Circuits and Systems I*, vol. 44, no. 2, pp. 130–142, 1997.

[Cha08] T.-H. Chang and L.-R. Dung, "Fourth-order cascaded $\Sigma\Delta$ modulator using tri-level quantization and bandpass noise shaping for broadband telecommunication applications", *IEEE Transactions on Circuits and Systems I*, vol. 55, no. 6, pp. 1722–1732, 2008.

[Cho07] H.-C. Chow and J.-Y. Wang, "High CMRR instrumentation amplifier for biomedical applications", in: *Proceedings of the International Symposium on Signal Processing and Its Applications*, pp. 1–4, Sharjah, United Arab Emirates, February 2007.

[Com04] E. Compagne, S. Maulet, and S. Genevey, "An analog front-end for remote sensor applications with high input common-mode rejection including a 16 bit $\Sigma\Delta$ ADC in 0.35 μm 3.3 V CMOS process", in: *Proceedings of the European Solid-State Circuits Conference*, pp. 459–462, Leuven, Belgium, September 2004.

[Cor05] M. Corradin, G. Spiazzi, and S. Buso, "Effects of radio frequency interference in OPAMP differential input stages", in: *Proceedings of the International Symposium on Electromagnetic Compatibility*, vol. 3, pp. 866–871, Chicago, USA, August 2005.

[Cro03] G. Croce, A. Andreini, L. Cerati, G. Meneghesso, and L. Sponton, "ESD in smart power processes", in: *Proceedings of the 12th Workshop on the Advances in Analog Circuit Design*, pp. 169–206, Graz, Austria, April 2003.

[Cur32] L. E. Curtis, "Electrical interference in motor car receivers", in: *Proceedings of the Institute of Radio Engineers*, vol. 20, no. 4, pp. 674–688, April 1932.

[Deu03a] B. Deutschmann, "Improvement of system robustness through EMC optimization", in: *Proceedings of the 12th Workshop on the Advances in Analog Circuit Design*, pp. 227–242, Graz, Austria, April 2003.

[Deu03b] B. Deutschmann and T. Ostermann, "CMOS output drivers with reduced ground bounce and electromagnetic emission", in: *Proceedings of the European Solid-State Circuits Conference*, pp. 537–540, Estoril, Portugal, September 2003.

[Dob03] P. Dobrovolny, G. Vandersteen, P. Wambacq, and S. Donnay, "Analysis and compact behavioral modeling of nonlinear distortion in analog communication circuits", *IEEE Transactions on Computer-Aided Design of Integrated Circuits and Systems*, vol. 22, no. 9, pp. 1215–1227, 2003.

[DPI] *Measurement of E/M Immunity 150 kHz to 1 GHz-Part 4: Direct RF Power Injection Method*, IEC 62132-4, Ed.1: ICs, 2004.

[Dut97] K. Dutton, S. Thompson, and B. Barraclough, *The Art of Control Engineering*, Addison Wesley/Longman, Reading, 1997.

[Enz96] C. C. Enz and G. C. Temes, "Circuit techniques for reducing the effects of op-amp imperfections: autozeroing, correlated double sampling, and chopper stabilization", *Proceedings of the IEEE*, vol. 84, no. 11, pp. 1584–1614, 1996.

[Far99] M. Farsi, K. Ratcliff, and M. Barbosa, "An overview of controller area network", *Computing and Control Engineering Journal*, vol. 10, no. 3, pp. 113–120, 1999.

[Fio01] F. Fiori, S. Benelli, G. Gaidano, and V. Pozzolo "Investigation, on VLSI's input ports susceptibility to conducted RF interference", in: *IEEE International Symposium on Electromagnetic Compatibility*, pp. 326–329, August 1997.

[Fio01] F. Fiori, "Operational amplifier input stage robust to EMI", *IEE Electronic Letters*, vol. 37, no. 15, pp. 930–931, 2001.

[Fio02a] F. Fiori and P. S. Crovetti, "Nonlinear effects of radio-frequency interference in operational amplifiers", *IEEE Transactions on Circuits and Systems I*, vol. 49, no. 3, pp. 367–372, 2002.

[Fio02b] F. Fiori and P. S. Crovetti, "Complementary differential pair with high immunity to RFI", *IEE Electronics Letters*, vol. 38, no. 25, pp. 1663–1664, 2002.

[Fio03] F. Fiori and P. S. Crovetti, "Prediction of EMI effects in operational amplifiers by a two-input Volterra series model", *IEE Proceedings on Circuits, Devices and Systems*, vol. 150, no. 3, pp. 185–193, 2003.

[Fio04] F. Fiori and P. S. Crovetti, "Investigation on RFI effects in bandgap voltage references", *Microelectronics Journal*, vol. 35, no. 6, pp. 557–561, 2004.

[Fio06] F. Fiori, "On the susceptibility of analog circuits to EMI", in: *Proceedings of the Advances in Analog Circuit Design (AACD)*, Maastricht, the Netherlands, April 2006.

[For79] M. L. Forcier and R. E. Richardson, "Microwave-rectification RFI response in field-effect transistors", *IEEE Transactions on Electromagnetic Compatibility*, vol. 21, no. 4, pp. 312–315, 1979.

[Gar98] F. Garcia, P. Coll, and D. Anvergne, "Design of a slew rate controlled output buffer", in: *Proceedings of the IEEE International ASIC Conference*, pp. 147-150, Rochester, USA, September 1998.

[Gie93] G. G. E. Gielen and W. M. C. Sansen, "Modeling of the power-supply interactions of CMOS operational amplifiers using symbolic computation", in: *Proceedings of the IEEE International Symposium on Circuits and Systems*, vol. 2, pp. 1381–1384, May 1993.

[Gie06] G. G. E. Gielen, "Design methodologies and tools for circuit design in CMOS nanometer technologies", in: *Proceedings of the European Solid-State Circuits Conference*, pp. 21–32, Montreux, Switzerland, September 2006.

[Giu06] G. Giustolisi and G. Palumbo, "NMOS low drop-out regulator with dynamic biasing", in: *Proceedings of the IEEE International Conference on Electronics, Circuits and Systems (ICECS)*, pp. 204–207, December 2006.

[Gla07] U. Glaser, *Complex ESD Protection Elements and Issues in Decananometre CMOS Technologies*, Hartung-Gorre Verlag Konstanz, 2007.

[Gly96] J. Glyn, *Modern Engineering Mathematics*, 2nd edn., Addison-Wesley, Reading, 1996.

[Goe85] J. J. Goedbloed, K. Riemens, and A. J. Stienstra, "Increasing the RFI immunity of amplifiers with negative feedback", in: *Proceedings of the 5th Symposium on Electromagnetic Compatibility*, pp. 471–476, 1985.

[Goe01] J. J. Goedbloed, *Electromagnetic Compatibility*, Kluwer, Amsterdam, 2001.

[Gra97] S. Graffi, G. Masetti, and A. Piovaccari, "Criteria to reduce failures induced from conveyed electromagnetic interferences on CMOS operational amplifiers", *Microelectronics and Reliability*, vol. 37, no. 1, pp. 95–113, 1997.

[Gra01] P. R. Gray, P. J. Hurst, S. H. Lewis, and R. G. Meyer, *Analysis and Design of Analog Integrated Circuits*, Wiley, New York, 2001.

[Gro02] D. J. Groot Boerle, "EMC and functional safety, impact of IEC 61000-1-2" in: *Proceedings of the IEEE International Symposium on Electromagnetic Compatibility*, vol. 1, pp. 353–358, Minneapolis, USA, August 2002.

[Gup05] V. Gupta and G. A. Rincon-Mora, "A low dropout, CMOS regulator with high PSR over wideband frequencies", in: *Proceedings of the IEEE International Symposium on Circuits and Systems*, pp. V-4245–V-4248, Atlanta, USA, May 2005.

[Has00] E. Haseloff, Texas Instruments, Application Report SLYA014A, *Latch-up, ESD, and Other Phenomena*, Available online: http://focus.ti.com/lit/an/slya014a/slya014a.pdf, May 2000.

[Has01] A. Hastings, *The Art of Analog Layout*, Prentice Hall, New York, 2001.

[Hea95] M. A. Heald and J. B. Marion, *Classical Electromagnetic Radiation*, Thomson Learning, 1995.

[Hil64] D. F. Hilbiber, "A new semiconductor voltage standard", in: *International Solid-State Circuits Conference, Digest of Technical Papers*, pp. 32–33, Philadelphia, USA, February 1964.

[Hyv03] S. Hyvonen, S. Joshi, and E. Rosenbaum, "A cancellation technique to provide ESD protection for multi-GHz RF inputs", *IEE Electronics Letters*, vol. 39, no. 3, pp. 284–286, 2003.

[Ida04] N. Ida, *Engineering Electromagnetics*, 2nd edn., Springer, Berlin, 2004.

[IEC 61967] IEC 61967, *Integrated Circuits, Measurement of Electromagnetic Emission up to 1 GHz*. Available online: http://www.iec.ch.

[IEC 62132] IEC 62132, *Integrated Circuits, Measurement of Electromagnetic Immunity up to 1 GHz*. Available online: http://www.iec.ch.

[IET] IEE, *IEE Guidance Document on EMC and Functional Safety*, Available online: http://www.theiet.org/factfiles/emc/.

[IEV] IEV, *International Electrotechnical Vocabulary*, Available online: http://www.electropedia.org.

[Ise02] R. Isermann, R. Schwarz, and S. Stolzl, "Fault-tolerant drive-by-wire systems", *IEEE Control Systems Magazine*, vol. 22, no. 5, pp. 64–81, 2002.

[ITRS] ITRS, http://public.itrs.net.

[Joh88] M. G. Johnson and E. L. Hudson, "A variable delay line PLL for CPU-coprocessor synchronization", *IEEE Journal of Solid-State Circuits*, vol. 23, no. 5, pp. 1218–1223, 1988.

[Kay07] M. Kayal, "CMOS linear regulators, design and case studies", in: *Electronics Laboratories Advanced Engineering Course on Power Management*, Lausanne, Switzerland, September 2007.

[Kei87] B. Keiser, *Principles of Electromagnetic Compatibility*, 3rd edn., Artech House, Norwood, 1987.

[Kon94] T. Konefal and A. C. Marvin, "Prediction and measurement of EMC radiated immunity problems in interconnected digital electronic systems", *IEE Proceedings on Science, Measurement and Technology*, vol. 141, no. 6, pp. 464–470, 1994.

[Kov00] Z. M. Kovacs-Vajna, E. Sardini, and N. Speciale, "Chaotic behavior of 741 opamps subjected to EMI conveyed to power supply rails", in: *Proceedings of the IEEE International Symposium on Circuits and Systems*, pp. I-727–I-730, Geneva, Switzerland, May 2000.

[Kui73] K. E. Kuijk, "A precision reference voltage source", *IEEE Journal of Solid-State Circuits*, vol. 8, no. 3, pp. 222–226, 1973.

[Lak94] K. R. Laker and W. M. C. Sansen, *Design of Analog Integrated Circuits and Systems*, McGraw-Hill, New York, 1994.

[Lau95] J.-J. Laurin, S. G. Zaky, and K. G. Balmain, "Prediction of delay induced by in band RFI in CMOS inverters", *IEEE Transactions on Electromagnetic Compatibility*, vol. 37, no. 2, pp. 167–174, 1995.

[Ler01] P. Leroux and M. Steyaert, "High-performance 5.2 GHz LNA with on-chip inductor to provide ESD protection", *IEE Electronics Letters*, vol. 37, no. 7, pp. 467–469, 2001.

[Ler03] P. Leroux and M. Steyaert, "RF-ESD co-design for high performance CMOS LNAs", in: *Proceedings of the 12th Workshop on the Advances in Analog Circuit Design*, pp. 207–226, Graz, Austria, April 2003.

[Ler05] P. Leroux and M. Steyaert, *LNA-ESD Co-Design for Fully Integrated CMOS Wireless Receivers*, Springer, Berlin, 2005.

[LIN] LIN Consortium, *LIN Specification Package Revision 2.0*. Available online: http://www.lin-subbus.org.

[Liv92] D. L. Livesey, *Introduction to Physics*, Wuerz Publishing, 1992.

[Loc04] C. Lochot, S. Calvet, S. Ben Dhia, and E. Sicard, "REGINA test mask: research on EMC guidelines for integrated automotive circuits", *Microelectronics Journal*, vol. 35, no. 6, pp. 509–523, 2004.

[Loe06] J. Loeckx and G. Gielen, "Efficient identification of major contributions to EMI-induced rectification effects in analog automotive circuits", in: *Proceedings of the International Zurich Symposium on Electromagnetic Compatibility*, pp. 148–151, Singapore, February 2006.

[Loe07] J. Loeckx and G. Gielen, "Assessment of the DPI standard for immunity simulation of integrated circuits", in: *Proceedings of the IEEE International Symposium on Electromagnetic Compatibility*, pp. 1–5, Hawai, USA, July 2007.

[Man83] T. Mano, J. Yamada, J. Inoue, and S. Nakajima, "Circuit techniques for a VLSI memory", *IEEE Journal of Solid-State Circuits*, vol. 18, no. 5, pp. 463–470, 1983.

[Mar88] A. C. Marvin, "Introduction to EMC phenomena", in: *IEE Colloquium on Vehicle Electromagnetic Compatibility*, pp. 1–7, March 1988.

[Mas96] G. Masetti, S. Graffi, D. Golzio, and Z. M. Kovacs-Vajna, "Failures induced on analog integrated circuits by conveyed electromagnetic interferences: a review", *Microelectronics and Reliability*, vol. 36, no. 7/8, pp. 955–972, 1996.

[Mas99] G. Masetti, G. Setti, and N. Speciale, "On the key role of parasitic capacitances in the determination of susceptibility to EMI of integrated operational amplifiers", in: *Proceedings of EMC Zurich*, pp. 625–630, 1999.

[Max01] Maxim, Application Note 883, *Improved Power Supply Rejection for IC Linear Regulators*, Available online: http://www.maxim-ic.com/appnotes.cfm/appnote_number/883, October 2001.

[Mey03] M. De Mey, "Robustness in analog design", in: *Proceedings of the 12th Workshop on the Advances in Analog Circuit Design*, pp. 243–253, Graz, Austria, April 2003.

[Mil07] R. J. Milliken, J. Silva-Martinez, and E. Sanchez-Sinencio, "Full on-chip CMOS low-dropout voltage regulator", *IEEE Transactions on Circuits and Systems I*, vol. 54, no. 9, pp. 1879–1890, 2007.

[Moe02] P. Moens, D. Bolognesi, L. Delobel, D. Villanueva, H. Hakim, S. C. Trinh, K. Reynders, F. De Pestel, A. Lowe, E. De Backer, G. Van Herzeele, and M. Tack, "I3T80: a 0.35 μm based system-on-chip technology for 42 V battery automotive applications", in: *Proceedings of the International Symposium on Power Semiconductor Devices and ICs*, pp. 225–228, Santa Fe, USA, June 2002.

[Mur03] T. Murakami, Y. Matsumoto, K. Fujii, and Y. Yamanaka, "Effects of multi-path propagation on microwave oven interference in wireless systems", in: *IEEE International Symposium on Electromagnetic Compatibility*, vol. 2, pp. 749–752, Istanbul, Turkey, May 2003.

[Ori08] E. Orietti, N. Montemezzo, S. Buso, G. Meneghesso, A. Neviani, and G. Spiazzi, "Reducing the EMI susceptibility of a Kuijk Bandgap", *IEEE Transactions on Electromagnetic Compatibility*, vol. 50, no. 4, pp. 876–886, 2008.

[Pap99] Y. E. Papananos, *Radio-Frequency Microelectronic Circuits for Telecommunication Applications*, Kluwer Academic, Amsterdam, 1999.

[Pat05] V. Patel and M. Steffka, "Vehicular spark ignition systems radiated emissions and reception performance", in: *IEEE International Symposium on Electromagnetic Compatibility*, vol. 1, pp. 19–23, Chicago, USA, August 2005.

[Pen68] A. A. Penzias, "Measurement of cosmic microwave background radiation", *IEEE Transactions on Microwave Theory and Techniques*, vol. 16, no. 9, pp. 608–611, 1968.

[Per07] R. T. Perry, S. H. Lewis, A. P. Brokaw, and T. R. Viswanathan, "A 1.4 V supply CMOS fractional bandgap reference", *IEEE Journal of Solid-State Circuits*, vol. 42, no. 10, pp. 2180–2186, 2007.

[Pes03] F. De Pestel, P. Coppens, H. De Vleeschouwer, P. Colson, S. Boonen, T. Colpaert, P. Moens, D. Bolognesi, G. Coudenys, and M. Tack, "Deep trench isolation for a 50 V 0.35 μm based smart power technology", in: *Proceedings of the European Solid-State Device Research*, pp. 191–194, Estoril, Portugal, September 2003.

[Phi80] V. J. Phillips, *Early Radio Wave Detectors*, IEEE & P. Peregrinus, 1980.

[Pou94] A. S. Poulton, "Effect of conducted EMI on the DC performance of operational amplifiers", *IEE Electronics Letters*, vol. 30, no. 4, pp. 282–284, 1994.

[Pre06] A. Pretelli, A. Richelli, and Z. M. Kovacs-Vajna, "Reduction of EMI susceptibility in CMOS bandgap reference circuits", *IEEE Transactions on Electromagnetic Compatibility*, vol. 48, no. 4, pp. 760–765, 2006.

[Ram09] M. Ramdani, E. Sicard, A. Boyer, S. Ben Dhia, J. J. Whalen, T. H. Hubing, M. Coenen, and O. Wada, "The electromagnetic compatibility of integrated circuits-past, present, and future", *IEEE Transactions on Electromagnetic Compatibility*, vol. 51, no. 1, pp. 78–100, 2009.

[Raz01] B. Razavi, *Design of Analog CMOS Integrated Circuits*, McGraw-Hill, New York, 2001.

[Red05] J.-M. Redouté and M. Steyaert, "Current mirror structure insensitive to conducted EMI", *IEE Electronics Letters*, vol. 41, no. 21, pp. 1145–1146, 2005.

[Red06a] J.-M. Redouté and M. Steyaert, "Improved EMI filtering current mirror structure requiring reduced capacitance", *IEE Electronics Letters*, vol. 42, no. 10, pp. 560–561, 2006.

[Red06b] J.-M. Redouté and M. Steyaert, "An EMI resisting LIN driver", in: *Proceedings of the European Solid-State Circuits Conference*, pp. 580-583, Montreux, Switzerland, September 2006.

[Red06c] J.-M. Redouté and M. Steyaert, "EMI resisting CMOS differential pair structure", *IEE Electronics Letters*, vol. 42, no. 21, pp. 1217–1218, 2006.

[Red07a] J.-M. Redouté and M. Steyaert, "An EMI resisting LIN driver in 0.35-micron high-voltage CMOS", *IEEE Journal of Solid-State Circuits*, vol. 42, no. 7, pp. 1574–1582, 2007.

[Red07b] J.-M. Redouté and M. Steyaert, "Electromagnetic susceptibility of CMOS analog integrated output structures", in: *Proceedings of the 6th International Worshop on Electromagnetic Compatibility of Integrated Circuits*, pp. 253–258, Torino, Italy, November 2007.

[Red07c] J.-M. Redouté and M. Steyaert, "An integrated CMOS differential pair immune to EMI", in: *Proceedings of the 6th International Worshop on Electromagnetic Compatibility of Integrated Circuits*, pp. 185–190, Torino, Italy, November 2007.

[Red07d] J.-M. Redouté and M. Steyaert, "Measurement of EMI induced input offset voltage of an operational amplifier", *IEE Electronics Letters*, vol. 43, no. 20, pp. 1088–1090, 2007.

[Red08a] J.-M. Redouté, C. Walravens, S. Van Winckel, and M. Steyaert, "An externally trimmed integrated DC current regulator insensitive to conducted EMI", *IEEE Transactions on Electromagnetic Compatibility*, vol. 50, no. 1, pp. 63–70, 2008.

[Red08b] J.-M. Redouté and M. Steyaert, "Active load for differential amplifier with high output impedance and reduced supply voltage", *IEE Electronics Letters*, vol. 44, no. 2, pp. 67–68, 2008.

[Red08c] J.-M. Redouté and M. Steyaert, "A CMOS source-buffered differential input stage with high EMI suppression", in: *Proceedings of the European Solid-State Circuits Conference*, pp. 318–321, Edinburgh, Great Britain, September 2008.

[Rib84] D. B. Ribner and M. A. Copeland, "Design techniques for cascoded CMOS op amps with improved PSRR and common-mode input range", *IEEE Journal of Solid-State Circuits*, vol. 19, no. 6, pp. 919–925, 1984.

[Ric79a] R. E. Richardson, "Modeling of low-level rectification RFI in bipolar circuitry", *IEEE Transactions on Electromagnetic Compatibility*, vol. 21, no. 4, pp. 307–311, 1979.

[Ric79b] R. E. Richardson, "Quiescent operating point shift in bipolar transistors with AC excitation", *IEEE Journal of Solid-State Circuits*, vol. 14, no. 6, pp. 1087–1094, 1979.

[Ric01] A. Richelli, L. Colalongo, Z. M. Kovacs-Vajna, and M. Quarantelli, "Design of an integrated CMOS operational amplifier with low probability EMI induced failures", in: *Proceedings of the European Solid-State Circuits Conference*, pp. 361–364, Villach, Austria, September 2001.

[Ric03] A. Richelli, L. Colalongo, and Z. Kovacs-Vajna, "Increasing the immunity to electromagnetic interferences of CMOS opamps", *IEEE Transansactions on Reliability*, vol. 52, no. 3, pp. 349–353, 2003.

[Rin98] G. A. Rincon-Mora and P. E. Allen, "A low-voltage, low quiescent current, low drop-out regulator", *IEEE Journal of Solid-State Circuits*, vol. 33, no. 1, pp. 36–44, 1998.

[San99] W. M. C. Sansen, "Distortion in elementary transistor circuits", *IEEE Transactions on Circuits and Systems II*, vol. 46, no. 3, pp. 315–325, 1999.

[San06] W. M. C. Sansen, *Analog Design Essentials*, Springer, Berlin, 2006.

[Sch80] M. Schetzen, *The Volterra and Wiener Theories of Nonlinear Systems*, Wiley, New York, 1980.

[Sch00] R. Schmitt, "Understanding electromagnetic fields and antenna radiation takes (almost) no math", *EDN Magazine*, pp. 77–88 , March 2000.

[Sch02] R. Schmitt, *Electromagnetics Explained: A Handbook for Wireless/RF, EMC, and High-Speed Electronics*, Newnes, London, 2002.

[Seg01] B. Segal, D. Davis, C. W. Trueman, and T. J. F. Pavlasek, "Risk of patient injury due to electromagnetic-interference malfunctions: estimation and minimization", in: *Proceedings of the IEEE International EMC Symposium*, vol. 2, pp. 1308–1312, Montréal, Canada, August 2001.

[Sen93] R. Senthinathan and J. L. Prince, "Application specific CMOS output driver circuit design techniques to reduce simultaneous switching noise", *IEEE Journal of Solid-State Circuits*, vol. 28, no. 12, pp. 1383–1388, 1993.

[Set97] L. Setian, *Practical Communication Antennas with Wireless Applications*, Prentice Hall, New York, 1997.

[Shu54] L. Shu-hua, "Origine de la boussole: II. Aimant et boussole", *Isis*, vol. 45, no. 2, pp. 175–196, 1954.

[Sic07a] E. Sicard, S. Ben Dhia, M. Ramdani, and T. Hubing, "EMC of integrated circuits: a historical review", in: *Proceedings of IEEE International Symposium on Electromagnetic Compatibility*, pp. 1–4, Honolulu, USA, July 2007.

[Sic07b] E. Sicard, L. Courau, and M. Fer, "Package modeling for EMC prediction: a 64 pin BGA case study", in: *Proceedings of the 6th International Worshop on Electromagnetic Compatibility of Integrated Circuits*, pp. 77–80, Torino, Italy, November 2007.

[Sic07c] E. Sicard, S. Ben Dhia, M. Ramdani, J. Catrysse, and M. Coenen, "Towards an EMC roadmap for integrated circuits", in: *Proceedings of the 6th International Worshop on Electromagnetic Compatibility of Integrated Circuits*, pp. 261–264, Torino, Italy, November 2007.

[Son83] B. S. Song and P. R. Gray, "A precision curvature-compensated CMOS bandgap reference", *IEEE Journal of Solid-State Circuits*, vol. 18, no. 6, pp. 634–643, 1983.

[Spe77] J. D. Spencer and D. E. Pippenger, *The Voltage Regulator Handbook*, Texas Instruments Inc., 1977.

[Spe99] N. Speciale and G. Setti, "Failures induced on bipolar operational amplifiers from electromagnetic interferences conducted on the power supply rails", in: *Proceedings of the Annual International Reliability Physics Symposium*, pp. 333–336, San Diego, USA, March 1999.

[Spi74] M. R. Spiegel, *Schaum's Outline of Theory and Problems of Fourier Analysis: With Applications to Boundary Value Problems*, McGraw-Hill, New York, 1974

[Spi94] M. R. Spiegel, *Schaum's Outline of Theory and Problems of Laplace Transforms*, McGraw-Hill, New York, 1994.

[Ste83] M. Steer, P. J. Khan, and R. S. Tucker, "Relationship between Volterra series and generalized power series", *IEEE Proceedings*, vol. 71, no. 12, 1983.

[Ste87] M. S. J. Steyaert and W. M. C. Sansen, "A micropower low-noise monolithic instrumentation amplifier for medical purposes", *IEEE Journal of Solid-State Circuits*, vol. 22, no. 6, pp. 1163–1168, 1987.

[Ste90] M. S. J. Steyaert and W. M. C. Sansen, "Power supply rejection ratio in operational transconductance amplifiers", *IEEE Transactions on Circuits and Systems*, vol. 37, no. 9, pp. 1077–1084, 1990.

[Ste91] M. S. J. Steyaert, "High frequency power supply rejection in analog circuits", in: *Proceedings of the Annual IEEE International ASIC Conference and Exhibit*, pp. T13-3.1/2, September 1991.

[Swa94] A. J. G. Swainson, "Another look at electromagnetic shielding", in: *International Conference on Electromagnetic Compatibility*, pp. 106–110, September 1994.

[Tha95] K. M. Tham and K. Nagaraj, "A low supply voltage high PSRR voltage reference in CMOS process", *IEEE Journal of Solid-State Circuits*, vol. 30, no. 5, pp. 586–590, 1996.

[Tro79] J. G. Tront, J. J. Whalen, C. E. Larson, and J. M. Roe, "Computer-aided analysis of RFI effects in operational amplifiers", *IEEE Transactions on Electromagnetic Compatibility*, vol. 21, no. 4, pp. 297–306, 1979.

[Tro85] J. G. Tront, "Predicting URF upset of MOSFET digital IC's", *IEEE Transactions on Electromagnetic Compatibility*, vol. 27, no. 2, pp. 64–69, 1985.

[Wal07] C. Walravens, S. Van Winckel, J.-M. Redouté, and M. Steyaert, "Efficient reduction of electromagnetic interference effects in operational amplifiers", *IEE Electronics Letters*, vol. 43, no. 2, pp. 84–85, 2007.

[Wam98] P. Wambacq and W. Sansen, *Distortion Analysis of Analog Integrated Circuits*, Kluwer Academic, Amsterdam, 1998.

[Wan05] J. Wang, X. Wang, X. Zhai, H. Wang, and B. Lampe, "CAN/LIN hybrid network for automobile", in: *Proceedings of IEEE International Conference on Vehicular Electronics and Safety*, pp. 348–352, Xián, China, October 2005.

[Wei03] G. Weightman, *Signor's Marconi's Magic Box*, Da Capo press, 2003.

[Wes01] D. A. Weston, *Electromagnetic Compatibility: Principles and Applications*, 2nd edn., Marcel Dekker, New York, 2006.

[Whi60] Sr. E. Whittaker, *A History of the Theories of Aether and Electricity*, Harper and Brothers, New York, 1960.

[Wid71] R. J. Widlar, "New developments in IC voltage regulators", *IEEE Journal of Solid-State Circuits*, vol. 6, no. 1, pp. 2–7, 1971.

[Wie06] A. Wieers and H. Casier, "Methodology and case study for high immunity automotive design", in: *Proceedings of the 15th Workshop on the Advances in Analog Circuit Design*, Maastricht, The Netherlands, April 2006.

[Wik00] M. W. Wik, W. A. Radasky, and R. L. Gardner, "The threat of intentional electromagnetic interference", in: *Proceedings of the Asia-Pacific Conference on Environmental Electromagnetics*, pp. 17–19, Shanghai, China, May 2000.

[Wu94] W. C. S. Wu, W. J. Helms, J. A. Kuhn, and B. E. Byrkett, "Digital-compatible high-performance operational amplifier with rail-to-rail input and output ranges", *IEEE Journal of Solid-State Circuits*, vol. 29, no. 1, pp. 63–66, 1994.

[Yav05] M. Yavari, N. Maghari, and O. Shoaei, "An accurate analysis of slew rate for two-stage CMOS opamps", *IEEE Transactions on Circuits and Systems II*, vol. 52, no. 3, pp. 164–167, 2005.

[Zin00] Y. Zinzius, E. Lauwers, G. Gielen, and W. Sansen, "Evaluation of the substrate noise effect on analog circuits in mixed-signal designs", *Proceedings of the Southwest Symposium on Mixed-Signal Design (SSMSD)*, pp. 131–134, San Diego, USA, February 2000.

[Zho08] J. Zhou and W. Dehaene, “Evaluation of power supply noise in CMOS and low noise logic cells”, in: *Asia-Pacific Symposium on Electromagnetic Compatibility and International Zurich Symposium on Electromagnetic Compatibility*, pp. 12–15, Singapore, May 2008.

[Zoe02] D. Van Zoest and D. Otten, “Implementation of a simulated washing machine with an integrated control panel and intra-appliance local interconnect network bus”, in: *IATC, 53rd Annual Conference*, 2002.

[Zve67] A. I. Zverev, *Handbook of Filter Synthesis*, Wiley, New York, 1967.

# Index




J.-M. Redouté, M. Steyaert, *EMC of Analog Integrated Circuits*,
Analog Circuits and Signal Processing,
DOI 10.1007/978-90-481-3230-0,

MIX
Papier aus verantwortungsvollen Quellen
Paper from responsible sources
FSC® C105338

If you have any concerns about our products,
you can contact us on
**ProductSafety@springernature.com**

In case Publisher is established outside the EU,
the EU authorized representative is:
**Springer Nature Customer Service Center GmbH**
**Europaplatz 3, 69115 Heidelberg, Germany**

Printed by Libri Plureos GmbH
in Hamburg, Germany